FORTY YEARS WITH FORD TRACTORS

David Pearson
with Martin Rickatson

Old Pond PUBLISHING

First published 2017

Book layout by Mark Paterson
Photos by:
1–11, 24–26, 31–33, 40 David Pearson collection
12–23, 27–30, 34–36, 38–39 Ford Motor Company/
New Holland archives 37 Bühler Versatile

A catalogue record for this book is
available from the British Library

ISBN 978-1-910456-59-0

Fox Chapel Publishers International Ltd.
20-22 Wenlock Rd.
London N1 7GU, U.K.

www.oldpond.com

We are always looking for talented authors. To submit an idea,
please send a brief inquiry to acquisitions@foxchapelpublishing.com.

Printed in the USA

Contents

Dedication

To my good friend Tony Speakman
and to my daughter Nicky

Foreword by Rory Day

The basis of this book first took the form of a series of 'Memory Lane' articles run over a period of 14 months in *Classic Tractor* magazine. I'm sure I wasn't the only one who was enthralled by David Pearson's tales of his experiences with the tractor division of the Ford Motor Company during a remarkable career that spanned almost 40 years.

From field-testing prototype Fordson Dextas in the mid-1950s to having a substantial input into the design of the high-hp Ford 70 Series almost 40 years later, David truly is a man who can say he has 'been there, done that' and lived to tell the tale.

He never tires of talking about the old times, and was always only too willing to clear up the occasional queries that cropped up during his long running series of recollections and stories. For those who have never met the man, David is a true gentleman, with none of the airs and graces that are usually associated with those who have served as senior management in a large company. His work was always more than a job, and even in retirement his love of Ford tractors remains as strong as ever.

This book, which encompasses all those reminiscences and more, plus the photos that were used and many additional ones, serves as a fitting tribute to his contribution to agriculture, to tractor development, and to Ford tractors in particular.

Rory Day
Editor, Classic Tractor
July 2016

Foreword by Andrew Watson

The agricultural machinery business is a relatively small community of people with a passion for farming and an engineering bent. Over the years there has been a small number of key figures who have shaped the industry and have led from the front. David Pearson is one of that number.

His vision, allied to his knowledge of farming and engineering and his focus on listening to and understanding the needs of farmers and dealers, has made him one of the most successful and special people in this business, and we owe him a considerable debt of gratitude. Among the many legacies he has left is that as the father of high-powered tractors. It was his vision that led during the 1970s to the development of iconic models including the first turbocharged Ford, the 7000, and the legendary FW-30 and FW-60 articulated models.

David was an inspirational leader from the day he joined Ford's tractor business in 1955. At that time, Bill Batty was CEO, and everyone understood that the very highest standards of performance were a fundamental requirement of all who worked in the company. David was a prime proponent.

Those who worked with him knew that striving to exceed these standards and fulfil the needs of dealers and customers, thereby helping the company attain the best possible market share, guaranteed David's total support. Known by almost everyone in the UK agricultural industry, he set an enviable standard of commercial performance.

David played a key role as manager of the Ford Tractor Training Centre at Boreham House during the critical launch in 1964 of the Basildon-built 6X series of tractors. Later he became sales manager for the southern region of the UK in the immediate post-6X years. This was a time that called for the ultimate in dealer support and customer relations, an area in which David excelled by establishing strong personal relationships. At the same time he focused on fleet owners, visiting them and establishing a unique level of trust between manufacturer and key customers. He was also a powerful motivator of his sales team, never tolerating second rate results.

David rose still further to become general sales manager, and the business thrived under his leadership alongside that of executive director Geoff Tiplady, particularly after the successful integration during the mid-1980s of the New Holland harvesting equipment range, creating Ford New Holland. David's well-deserved retirement from the business coincided with the beginnings of the formation of today's New Holland tractor and harvesting brand.

He still lives within a stone's throw of Boreham House, not many miles from Basildon, and stays in touch with the business he helped to create. David is a true legend of our industry, and his drive and vision are inspirational to us even today.

Andrew Watson
Business director, New Holland UK and Republic of Ireland
Basildon, July 2016

1

Beginnings in Agriculture

When I was a small boy of eight, I discovered a farm at the end of the road along which we lived. Olders Farm, in Angmering, West Sussex, was 50 acres of mostly grassland, with two acres of fodder beet for the dairy unit of 20 Ayrshire cows. Fascinated by its comings and goings, I asked the farmer if he would allow me to help with mucking out and feeding. So began my fascination with farming.

With the promise that one day I would be paid, probably half a crown, I was allowed to take on certain farm tasks. I loved the work, and would happily spend every day of my school holidays there. Our only form of power was a Shire horse, which pulled the cart that carried the manure out to the field, but neither this nor my father's warning against my choice of industry could put me off – I wanted to work in farming.

As my youth progressed, I continued spending every spare moment on the farm. One Saturday, when I was about 16, I was invited to a wedding at the local golf club, but didn't plan a late evening – I had promised to look in on a sow due to farrow later that night. The best laid plans often go awry, though, and while at the reception I was introduced to Martinis. In my innocence, I had no idea they were neat alcohol, and while I did get back to check on the pig, all I can remember was waking up at 6 o'clock the following morning with a very sore head, surrounded by the sow and 21 piglets.

Two years later, I can recall my father telling me that if I was ever to get married and start a family, I would have to find a job that paid more than the £4 a week to which I had finally graduated. With my agricultural interests in mind, he suggested I write to Massey Ferguson and to Ford, the two biggest names in tractors at that time.

December 1954 saw me taking the trip up to Coventry for an interview at the MF factory. To my surprise, I was offered a job on the spot, and was told I could start work the following September. But that was nine months away – and I couldn't afford to wait that long.

My next trip was to Dagenham, home at that time of Ford's UK tractor manufacturing business, where the world-famous Fordson Major was produced. Somehow I got an appointment to see Harry Power, the head of the Fordson manufacturing operations, and father of Harry and John, who both later worked for the company. I explained to Mr Power that all I wanted to do was drive tractors, whereupon he told me that I would have to learn how to make them before I could do that!

Again, I was lucky enough to be offered a job on the spot, and this time it was to start almost immediately. I gratefully accepted, and was told I could begin work on the tractor line in D Building at 7.15am on Monday, 10 January 1955, on a salary of six pounds and three shillings a week. The parents of my father's secretary lived in a house in Upminster, which was a train and bus journey away from the plant, and initially looked as if it would provide handy lodgings. However, I omitted to tell them about the 7.15am start time, which meant I had to creep around when rising at 5.30am to catch the train to Dagenham Heathway and then a bus to the plant. I found this very embarrassing and soon found new digs with people who were more used to getting up at this ungodly hour.

On my first day at the plant, I found the line foreman and introduced myself. Ten minutes later I was straight in the thick of things, having been sent to D building, where engines were tested below ground in the dynamometer test beds. Each test bed had two engines, one running and one waiting its turn. There were ten bays, with two engines and one man in each bay, and each engine was run for one hour, in heat of almost 90°F – outside there was 10in of snow. If an engine was deemed to be running OK, it then started on its journey down the line to becoming a Fordson Major, the key Ford tractor model of the period.

I was told to dress in a white coat – although everyone else seemed to be wearing blue overalls. The engine tester, who was supposed to be supervising me, ignored me until, after two hours of watching, I asked if I could help connect up the inlet and exhaust, bolt the engine to the dynamometer and turn on the water for cooling. He looked at me in amazement but agreed, and after an hour he realised I had got the hang of it and sat down to read his paper. We were now mates. When I later returned from lunch he had a pair of blue overalls for me, well soaked in oil. He then removed my white coat and placed it in the bin, informing me I wouldn't be needing it any more. I later learned that white coats were worn by inspectors and a change of clothes made me 'one of the boys' instead of one of 'them'. This was undoubtedly a very important day in my life and never to be forgotten. I soon knew the first name of every man on the line, and to them I became 'Dave'.

After two weeks on the dynamometers, I moved on. Every other Monday morning I got a new job in a new location. By week three, I was repairing the engines that hadn't performed correctly on the dyno test. Slowly but surely I built up the engines and gearboxes, assembled rear transmissions and finally, at the end of the year, reached the end of the line – the repair floor. This is where tractors with problems or parts missing were repaired and approved for shipment.

Tractors going for export did not have batteries fitted, to prevent theft during shipping, so in order to start them batteries on trolleys were connected to the tractors' terminals. To start the tractor, the throttle was fully opened, the starter depressed, and the cable then connected from the battery. We were always short of space, and on one occasion there were three Majors lined up nose to tail. If I had read the red tag that one day had been fixed to a particular tractor, I would have known it was seized in first gear and I wouldn't have tried to start it. As a result we now had three tractors all needing new front hoods and bonnets! My mate on the repair floor, Alf, rushed over, removed the warning tag and complained to the foreman that I could have been killed …

Just as I was wondering what I would be doing the following week, a man I hadn't seen before came across to tell me that next Monday I should report to Mudlands Farm at Rainham, four miles away, better known as the field test site. All modifications to existing tractors were tested here, and prototype tractors started their life at Mudlands before being sent out to work on farms in the local area, which comprised very heavy land adjacent to the River Thames.

At the time I arrived, the engineers had been doing some re-engineering exercises on the Major's crown wheel and pinion to try to reduce production costs. Testing involved doing figures of eight, turning left with the left brake on and then vice versa. However, the new reduced-cost crown wheel and pinion seldom lasted long enough to get the tractor out of the shed, and we were changing four or five a day! In the end, the need for reliability won the day and we kept the original tried-and-tested arrangement.

As the tractors were on test 24 hours a day, everyone had to do some night work. On the edge of the River Thames where the test track was situated, large piles of soil had been dumped many years previously during the building of the London Underground, and it was quite a spooky place. One night driving round the track pulling a weighted sledge, I got a tap on the shoulder and turned to see a bald man with a stocking over his head. I have to admit I was terrified – until I realised it was a mate doing the same job as me who'd got bored and decided to pull a prank!

The manager of the test site was a wonderful man named George Smale, and I later worked at Ford with his son, Bob. We had a big shed that represented desert working conditions, with a floor of cement dust that also came to cover the walls and ceiling. One day a young lad was working in there and I was persuaded to throw a half-shaft on to the roof to dislodge the cement – as you can imagine, the air became choked with dust and it was impossible to see a thing. A full five minutes passed before he crawled out, choking and spluttering. It wasn't until then that the colleagues who'd goaded me into playing this prank told me the boy's father was the plant manager! Fortunately, no one else ever knew who the culprit was …

I had been promised that I would soon be driving the prototype of the new Dexta tractor, named not after the breed of little black cattle, but as in dextrous – skilful and able. In March 1956 a message came to me to cancel my digs and collect all my gear. Along with two charming Irishmen – one, Ted Lonergan, who became a very good friend – I was to then go with two new prototype Dextas and a little grey Ferguson for comparison to our test site, which was to be the Elveden Estate in Suffolk. The new task would see me back in farming for the next two years.

The digs Ford had chosen for us were in a pub sited just 6ft from the main railway line, and we were shown a double bed and told we were all sleeping in it! Before I had time to refuse, the two Irishmen had said they were not having it and walked out. Eventually, we found better lodgings at the Bell at Mildenhall, which was to be our home for the next two years. Daily expenses were 16 shillings – about 75 pence – which paid for our bed, breakfast, a packed lunch and an evening meal. We had jugs and bowls in the room for washing and shaving.

Elveden Estate is one of the best pheasant shoots in the country. King George V was a regular visitor to the shoot, but never drew a peg – he always had the King's Stand, right in the middle. Gamekeeper figures I was shown from the estate recorded a bag of 3,248 birds in one day in 1912, and 22,788 pheasants in the 1904–05 shooting season. And vermin control was always a high priority for the keepers; in 1903 alone there were 14,662 rats killed on the estate.

We arrived at Cranhouse Barn, which adjoined the house of the keeper, whose name was Lofty – he was well over 6ft tall. The farm was overrun with pheasants, and in those days I considered it a prize if I could chase one on to the verge and take it out with my car's front numberplate. The temptation was enormous, and although we were warned not to even think about it, I have to admit we did do a bit of poaching. Elveden grew huge fields of lucerne, and when we drove home at night across a field, hen pheasants roosting on the ground would take off at the sight of the tractor lights, not knowing where they were in the darkness. After a bit of practice we could actually catch them in our hands, or better still get them with a stick and go back and pick them up. I should add that this only ever took place on a Thursday night prior to going home for the weekend, never more than two each and only in the shooting season …

In later years, Ernest Doe, of the well-known Ford tractor dealership and a legend in his own lifetime, regularly took me shooting at Elveden. Over a drink after one shoot I admitted my youthful activities to Lofty the keeper, who was still working on the estate. All he said was: 'Tell me something I don't know …!'

Elveden Heath had been a battle training ground during the war, and we ploughed thousands of acres over the two years, almost every day picking up unexploded mortars or shells between plough mouldboards and discs. Fortunately for us, none of them ever exploded!

As a vast 25,000-acre estate of mainly light land, we could work in any weather. I had changed from a 45-hour week being paid £6.30 to an 84-hour week being paid for a 140-hour period of two seven-day weeks on and two days off. That was a lot of money in those days, but we couldn't think of a better place for it to go than the till behind the bar at the Bell.

It was about this time in my life, at just 21 years old, that I was presented with a golden opportunity for a young man. I was back home in Sussex, where the local golf club always had a big New Year's Eve ball, and I got a phone call from the secretary asking if I would like to partner a young Canadian girl staying at the club over Christmas. Hesitantly, I said yes.

She was a charming girl and we had a great evening, after which I invited her out for lunch the following day. I arrived at her hotel to pick her up in my mother's Ford Prefect, but she took one look at it and suggested going in her car, which to my amazement was a brand new Ford Zodiac drop-head. This should have told me something: her car was worth more than I earned in a year. We saw quite a lot of each other for the few days left before she returned to Canada, and I still have the stuffed tiger she left me as a gift. When a friend later suggested that I had 'missed out', he had to repeat her surname to me until I could work out what he meant – she was a Johnson, of the Johnson & Johnson pharmaceuticals family.

2

New Fordsons for a New Era

As 1957 came around, we had finished testing, and it was time to launch the new Dexta as a smaller companion for the Major. I was moved from Suffolk to Ford's training centre, the Boreham House stately home in Essex, to show our dealer salesmen what the new tractor could do. West Country dealers were delighted that the new Dexta would enable them to compete against the popular Massey Ferguson 35, while for those in the more arable east the Major remained the more important machine. The Dexta, though, was a brilliant tractor to demonstrate, with lots of torque. With a two furrow plough attached, it was possible to set the throttle at 800rpm and get off and walk beside the tractor – not something that 'health and safety' would approve of today! On very cold days in Elveden we often did this to keep warm, until one day one of the drivers, Wally Ward, slipped as he got on and went under the rear wheel. He escaped, black and blue with bruises – fortunately the ground was frosted, which saved his life.

If there was a spare bed at Boreham House, I used to stay the night during the week, but at the weekend the house was empty, and I had my pick of the rooms. Often I would find tasks to do to keep me entertained over the weekend, and one day I decided to clear the shingle banks that slowed the water flow of the brook that ran through the grounds. As I loved driving the County crawler with dozer blade that we had on loan, I began to use it to shift the stones. I'd not been in the brook very long, though, before the track drive wheels filled with shingle and I was stuck. My first thought was to call out Fred Ridout from Co-Partnership

Farms at the bottom of the drive. He arrived with a winch-equipped tractor, but it wouldn't move the County, and we soon had a snapped wire rope.

To my embarrassment, the crawler stayed stuck in the brook for a couple of weeks, with the water at seat level. The next person to try and help extract it was an ex-army man working for the company, who suggested this would be an excellent exercise for the tank recovery unit at Colchester Garrison. This, however, turned out to be overkill – about a dozen soldiers with officers arrived and started to study the situation. Their main power unit was a Thornycroft Antar 6 × 4 tractor unit, complete with miles of steel cable, sprags, chains and a trailer load of equipment. The Antar was so heavy it broke all the drain covers as it came down the drive. Meanwhile, the County had to come out forwards as there was a bridge immediately behind it, and in front was a spinney about 100 yards long.

It was decided that the Antar would go to the other end of the wood and pay out cable back to the front axle of the County, going under the blade. At first the Antar slid backwards and the County stood still. Sprags were fitted and slowly but surely the County started to move, with the tracks locked solid. Once it was out of the brook and back at the tractor yard, the soldiers were welcomed in for a VIP lunch – once they had removed their boots of course. I, though, was not invited …

After they had gone, Bill James, the Boreham House manager, called me into his office and asked me how I intended to pay for the damage to the County and the cost of calling out the army. I was speechless, although I did find the words to tell him I earned only £13 a week. He got very busy with a pencil and decided payment would take about five years! Still dumbstruck, I listened as he said we would have to work something out as five years was far too long. I was sent off and worried for a week before he told me County had agreed to repair the crawler free of charge. I don't think we ever had another County crawler at Boreham House.

At this time, we would still be teaching many dealer staff the basics of demonstrating a tractor, including backing two- and four-wheel trailers. As part of this, we held a competition to see who could produce the best time manoeuvring a tractor and trailer – a two-wheel and then a four-wheel – from one place to another. The idea was that you backed the trailer across the yard and then selected a fast forward gear and drove back into the shed to park up. Trying to show off my skills one day, I stuck the tractor in top gear without accounting for the wet weather and greasy yard, and after slamming on the brakes as I approached the shed, skidded the last 20ft and managing to clip the edge of a stack of drums of blue paint. I think only one drum went over, but if that shed is still there its floor could well still be coloured 'Ford Tractor Blue' to this day.

***

The launch of the Dexta to the dealer principals was held at London's Alexandra Palace in 1957, ahead of a full 1958 launch. The event began with a slide presentation in a blackened dome, which at the end of the presentation opened out on to the main hall where a turntable with a Dexta sat centre stage – and on it sat me, waving to the dealers. That same picture of me was reprinted 25 times in the Farmers Weekly issue that covered the launch, and as a result I became known as 'the Dexta boy'. There were ten more Dextas in the hall but none of them had gears in the gearboxes – not that this was evident to the dealers present. They were delighted to have another tractor to sell alongside the Major. The new tractor was priced at just £500, or £550 with a double clutch.

A year after the 1958 launch, a team of us went to the World Ploughing Match in Northern Ireland, where we had around 50 tractors on show, each with its own implement and doing one bout four times a day. A team of us spent days preparing the site, staying near the Giant's Causeway, at the Ballycastle Hotel, where there were some gorgeous receptionists. That night I had only been in bed ten minutes when a girl was screaming and hammering at my door. I opened it to be confronted by one of the aforementioned receptionists, to be told that she had a wasp in her room that needed killing. I gallantly did so and assumed she would go back to bed, but was more than mildly surprised when she made plain she was too terrified to return to her room and would be staying with me ...

The following day, with the ploughing match under way, unbeknown to me my colleague John Prentice had got talking to one of the Ransomes field men, Willie Guthrie, who, although ready for retirement, was well known as a Cumberland wrestler. John convinced him that I was also interested in wrestling, which was totally untrue, and pointed me out to Willie, who approached me saying he'd heard I wanted to wrestle! Before I had time to speak he had removed his hearing aid and grabbed hold of me, and I was on the ground screaming for help. With some difficulty, I explained to him that there had been a misunderstanding, after which he helped me up, we shook hands and carried on with the work of demonstrating – that was our only wrestling incident!

Of those who were ploughing with our tractors, I particularly recall Jean Burns, from the Isle of Man. She used to drive us all mad with her requests for changes to wheel widths, and even wanted the land wheel of her Fordson Major put on the wrong way round. She was, though, a very good ploughwoman.

At this point, my life underwent another big change. I was called into Dagenham and told that I was going to Central and South America to launch the Dexta to our dealer organisation, accompanied by Derek Barron, the sales representative for the area. Derek was a good Spanish speaker, and later became the director in charge of Ford's UK car business. Before I knew it I'd been handed a wad of plane tickets that would take me to New York, Cuba, the Dominican Republic, Jamaica, Trinidad, Colombia, Venezuela, Panama, Costa Rica and Guatemala. In those days everyone at Ford flew first class, and on the first leg of my journey at the age of 21 years I climbed aboard a BOAC DC-7 with a sleeper, which meant climbing into what is now the luggage rack.

Total flying time to New York was 14 hours, and on arrival I met up with Derek before we boarded another plane bound for Cuba. On the way down it got very hot on board, but as I could neither see out of the windows nor speak Spanish I had no idea that one of the engines had been on fire and was no longer working! We had dropped to fly at 5,000ft, and were hugging the coast in case we had to make an emergency landing. I now realised why a lot of passengers were getting very jumpy, but with the skill of the pilot we landed safely in Havana and a new adventure started.

The first few days were spent on the farm of a huge lager factory. When we arrived in the field a crowd of locals had already lit a fire and were threading strips of beef on to what looked like runner bean poles that were set up like a teepee. A whole rear end of beef was being cooked and marinated using a brush made from a bunch of spring onions. This was to be lunch for us and the girls who worked at the factory.

I had never used a disc plough before, and it was some time before I realised that the top link was the vital ingredient in making it work successfully, aiding the ground penetration of the rear disc. Having mastered the implement, it wasn't long before we had ploughed a couple of acres and it was lunchtime. It was certainly the best beef I had ever tasted and it was washed down with gallons of local lager.

Several days later we were disc ploughing sugar cane stumps at the eastern end of the island. I couldn't help but notice the constant noise of rifle shooting in the distance, and asked if there was a rifle range nearby. The response that the country was in the midst of a revolution was more than a little scary, and when we went into town for lunch and there were dead bodies in the street, I have to admit I was terrified. The next day we saw fields of cane ablaze as a result of being set alight by the terrorists. But we soon moved on, and the following day we headed west, with no more excitement of that nature.

Cuba's agriculture is dominated by sugar, and I recall watching that week as a cargo boat was being loaded in the docks. It was a slow, manual process – one man, one sack of sugar. I remember asking why an elevator wasn't used, but it was soon pointed out to me that this would put a lot of men out of work.

We then moved to Jamaica, where we found the country in the grip of a national strike, one of the results being that no new tractors were being shipped in. We moved on to Trinidad, planning to return at the end of the tour.

Here, our demonstrations again centred around disc ploughing, but what I remember best are the wonderful hotels and perfect weather. From there, on we went to Bogota in Colombia, 18,000ft above sea level. We simply landed straight on to the runway – there seemed to be no descent – and even walking upstairs was exhausting due to the lack of oxygen.

There was one other bit of excitement still in store for me when we got to our next stop, Costa Rica. A Dexta had been sold to a farmer in the mountains, and poor access meant the only way to deliver the tractor was to fly it in piece by piece. To start we roped the engine to the belly of the plane and took off, beginning another terrifying flying experience, but after four or five trips we had all the parts in place and the tractor was reassembled on the farm.

In the late 1950s a man earning £20 a week or £1,000 a year was considered well-off. In those days, when working overseas we got a £50 a month bonus, but I was hourly paid and therefore earning overtime, plus overseas allowance and generous expenses. To say I'd never had it so good was an understatement.

3

Overseas Adventures

Sir Patrick Hennessy, the-then chairman of Ford of Britain, had apparently sat next to Lord Beaverbrook at a dinner in London sometime in the late 1950s, and in conversation it had become apparent to Sir Patrick that Lord Beaverbrook owned a large number of Massey Ferguson tractors on his Somerset farms. In order to try and persuade him of the merits of Fordson tractors, the loan of a Dexta for a six-month trial was agreed, and it followed that an instruction came to me to go to Dagenham, pick up a demonstration Thames Trader truck, drive it to Boreham House, collect a Dexta and take it to Somerset.

I collected the truck, which to my relief was full of fuel. I was slightly concerned, though, when the gauge oddly still showed full as I reached the busy Dartford Tunnel on my way to the South Circular Road. As I reached the halfway point, the truck coughed and stopped. Horns were blowing, oncoming traffic was pointing at me, a bus tried to pass me, but all I could think about was the £25 fine for blocking the tunnel, which at the time represented about four weeks' wages.

Then a miracle happened. A low-loader overtook me and backed up to the front of my truck. The driver appeared with a bundle of ropes, and in a few minutes we were almost hooked up when a policeman appeared. He asked what was going on and the driver explained things away by implying that he had been towing me when his rope had broken. We would be moving as quickly as possible, he said, and luckily he was right. Crawling slowly out of the tunnel, he delivered me to the first fuel station we could find. The only money I had on me was pound notes – worth a lot in those days – so I gave one to the driver, who would probably have been delighted with a few shillings, and thanked him profusely. Given the size of his reward, he said he would happily come down from Scotland and help me out again if I paid that generously!

Ten minutes later I had a full tank, had bled the fuel system, made a dipstick for the fuel tank and recommenced my trip to Somerset. To say I'd been scared would have been an understatement, as in those days blocking the tunnel like that was a major incident. Thankfully, the rest of the trip was uneventful – although to the best of my knowledge Lord Beaverbrook never did buy a Dexta.

It was when I was working at Boreham House that I met the girl who was to become my wife. Mollie worked as the assistant housekeeper there, and was a superb cook. We were married on 13 December, 1958, just after that year's Smithfield Show, and bought our first house in Ingatestone, Essex, for £2,200, which seemed a fortune at the time. Our daughter, Nicky, who took down every note of this story for me, was later born there.

No sooner had I settled in, though, than I was told to organise another demonstration team, this time to go to Europe. Four Thames Traders were needed to carry all the gear, which included twenty-four 16ft high lamp posts, each equipped with two 1,000w lights, and a five-cylinder Gardner diesel engine with 50kW generator to power them. Then there was a radio-controlled Dexta, and a huge inflatable tent that was to be used as a dance hall in the evening.

The plan was to start in Belgium, and then go on to Holland, Sweden and Finland. Two or three Ford dealers in each country organised a central farm for the events, which lasted a whole day at each location, with demonstrations and driving during the daytime followed by dining and dancing in the tent in the evening.

All went well until one day in Holland when we forgot to refuel the generator, all the lights went out and the tent started to deflate. There was pandemonium as people got out knives to cut the skin of the tent and escape! Luckily, fuel was close by, the generator was refilled and very soon we were up and running again, helped by some spare patches we fortunately had for the tent.

We were moving from hotel to hotel and to a different demonstration site every other day, and at one hotel in particular there was an extremely attractive waitress who was very good at looking after us. One day when we returned to the hotel I found a red rose in my tooth mug above the wash basin and, on further inspection, a note under the glass that read: 'I finish work at 9pm – is it possible we meet in the churchyard? Elsa x.'

The spelling was terrible and the English wasn't very good so I assumed it must be genuine. At 8.45pm I excused myself from the others at the bar and walked across to the church. To my great distress, no one appeared at 9pm. I thought I would give it another half-hour, and then, in case something had gone wrong, I waited until 10pm. Eventually I returned to the bar, to be met by Dick Drijver, the tractor manager for Ford in Holland and a man who became a great friend.

'Didn't she come?' he asked, innocently. Everyone was highly amused, knowing I'd been waiting for an hour under false pretences, but I failed to see the funny side!

Not long afterwards, we said a sad goodbye to Dick and to Holland, a very Ford country, and took the road to Germany, a very anti-Ford country. There were no demonstrations planned, but we had to drive on and through Denmark to catch the ferry to Sweden.

I was driving the second truck and saw the first one slowing down, pouring oil all over the road. I stopped behind it, and my colleague Jimmy Mitson, who'd been driving, declared the engine to still be running but producing no power. On closer inspection we saw a hole in the block the size of a dinner plate and a piston lying on the ground in the middle of the road. This was a disaster – we had the trucks booked on the ferry in two days' time. I phoned Ford's office in Cologne, only to be told that they did not use the Dagenham engines in Germany, but that the Ford dealer adjacent to the ferry port in Denmark should be able to help us. That, though, was hundreds of miles away!

We had a fixed tow bar with us, and I agreed to tow as I didn't fancy steering a truck with no brakes, no power steering and no lights. We pressed on to the hotel that we'd booked in Hamburg, but didn't arrive until 2am. The other two trucks had gone ahead and everyone was tucked up in bed. We had no idea how to get out of Hamburg so we rang the police and in no time we had a police escort at 7am. They were very pleased to see us on our way.

We got to the Danish truck dealer late the next evening and he worked all night to get us on the ferry to Sweden the next morning. We didn't have any demonstrations there, but were en route to Finland. I was the lead lorry and was very pleased to give a ride to a Swedish girl thumbing a lift at the side of the road. I speeded on to get away from the other trucks but was thwarted by a low bridge, where the tilt, with its canvas top, got stuck. We had no alternative but to remove all three tilts (luckily, the generator truck didn't have one), go under the bridge, replace them and continue on our way. It perhaps wasn't a surprise that the girl soon got bored and hitched a new lift.

With its dirt roads, Finland was hard work for the Thames Traders. Everything came under stress, from the mirrors, which fell off, to the driveshaft, which failed. I remember the country for its forests and the fact that all we ever seemed to eat was pea soup, so I wasn't sorry to come home.

But no sooner was I back than I received a telegram asking me to pack my bags again and prepare to leave Dagenham at 8.30am the following morning to drive to Belgrade in Yugoslavia, in the company of zone manager, Derek Rogers. Fordson was to compete in a contest between tractors from Massey Ferguson, Zetor, the Yugoslavian maker Zadrugar and a couple of other Eastern Bloc manufacturers. The idea was that the winner would get the contract to build a factory in Belgrade to build tractors. However, what we didn't know was that Massey Ferguson had already got the contract ...

We drove flat out in Derek's Mk 1 Consul, with a very short night stop in Switzerland, and arrived in Belgrade at 6pm the next day, having covered 1,800 miles in 48 hours. There we found a highly organised test, with marshals measuring the exact depth of work when ploughing, cultivating and rotary cultivating. The first test was deep ploughing at 11–12in, for which we had a two-furrow Ransomes TS64 plough that refused to scour on our trial run. Somehow we managed to find a metal polisher, who worked all night on the boards to shine them for the following day's work.

As we were preparing to go down from our room to dinner that night, a rather strange thing occurred. There was a knock at the door and I opened it to find a very attractive girl holding a bottle of the national drink, Slivovitz plum brandy. Derek and I were convinced she was a spy – remember this was a communist country – but she told us she had heard our English voices through the wall and invited us to share a drink. It turned out to taste like petrol, but we politely said how nice it was while she chatted about Belgrade. She turned out to be harmless and clearly not a spy! Rather merry after the brandy, I recall going steadily down to dinner at the hotel.

The following morning the test started with deep ploughing, and the boards worked perfectly, allowing us to meet all the rules set by the men following with tape measures. The field we were ploughing was very rough, with a lot of stalky grass, and to my horror one of the lower link linchpins was pushed out and one side of the plough came off its lower link. I was able to get the plough back on the tractor but had no spare linchpin. Generous help from a competitor was at hand, though, with Chalky White of Massey Ferguson producing a spare pin to set me on my way again.

When the ploughing was over we were told to fit rotary cultivators. I was driving the Major and the man next to me was driving a Zadrugar, a Yugoslavian-built tractor that resembled an E27N. He got off the tractor to adjust the top link, but hadn't disengaged the power take-off, and with there being no PTO guard his trousers were quickly caught and ripped off him, leaving him standing stark naked in the middle of the field with a broad cut on his leg. I jumped off the Major and pulled the stop control to avoid any more damage being done.

At the yard where all the tractors were stored there was a row of unused huge Russian crawlers, all at least three years old and unmarked. They were all fitted with donkey engines to start them. I was very keen to try out one of these huge vehicles, which resembled tanks, but after hours of trying to start one of the donkey engines with no success, I gave up.

One evening while seeking out some after-hours entertainment in Belgrade, Massey Ferguson's Chalky White introduced us to the Twibsky Dom Club, for Englishmen abroad. It showed us a film each night, and it was possible to buy English beer, which made for a nice way to spend the evening. When the film had finished they always played God Save the Queen, but I recall one night when Derek had fallen asleep and slept right through the anthem, much to the amusement of all the other guests.

He had to fly back to England the next day as his wedding was approaching rapidly, and I was not looking forward to my lengthy journey home alone. Some of the roads were hazardous, especially through Switzerland where my route took me along long stretches of single track road with 6ft of snow on either side. If you met a truck you had to back up until you found a passing space. The idea of staying at one of the luxurious Swiss hotels along the route was appealing, but I decided to press on and get home. Upon my return, though, life was about to change again.

***

I was next sent back to Boreham House as the senior sales lecturer, teaching our dealer organisation how to sell tractors, and I soon settled into my new role. It wasn't all work and no play, and the social side of being at Boreham was especially memorable – the Boreham House bar was famous with Ford staff, dealers and customers.

Thursday nights at Boreham House were party nights, because it was the last night of one-week courses, and most of the lecturing staff stayed to join in. Not a lot happened before 10 o'clock, but then my colleague Martin Brown would play the piano and anyone with a song in his head would sing. Boat races usually got started at 11pm, and involved two teams of people having a relay to drink half pints of bitter and return their mug to the table upside down. The staff were experts at this – they'd had lots of practice and I don't remember them ever being beaten. Mike Clarke was the king of the boat race game, and on many occasions he and I would take on a team of eight with half pints, while Mike and I had two pints each – and were never beaten.

We also had a glass yard that held two and a half pints. You had to be very precise in the way you held it if you were going to avoid spillage, and most people got more beer on the floor than they did in their mouths! Mike and I both took great pride in the fact we could drink three pints in 30 seconds.

More complex games included 'Cardinal Puff', where a good memory was required to recall a number of words. Anyone making a mistake had their glass refilled, and it was usually a student who invariably got worse and worse for wear. They never did get the better of us.

It was on one of these Thursday nights that Geoff Tiplady, who had not long been the manager of the tractor plant, was on his way home at around midnight. I lived in Ingatestone, while Geoff was a little further on, in Shenfield. We set off together with me following Geoff's car, and when I turned off to Ingatestone I imagined he would be OK to get the last three miles on his own. His wife was expecting a baby at time and to avoid waking her he parked his car down the road in a friend's drive, crept into his house and shut himself in the spare bedroom.

At 4am, his wife Sylvia went into labour and rang for an ambulance to take her to hospital. Still in the spare room, Geoff slept through all of this, his wife not knowing he was in the house just feet away. Desperate to know where he was, she rang the police – who in turn rang John Power, the duty manager at Boreham House – and then went off to hospital. John told them Geoff had gone home at the same time as me, and then at 5.30am rang me to explain the problem, whereupon I drove to Boreham House to meet John and decide what to do. We both went to Shenfield to look for Geoff, half expecting to find him in a ditch! We knew the road Geoff lived in but not the house number, although we then found his car in the friend's drive – and presumed we'd found his house. It was now about 7am, and we knocked on the door of the house where the car was parked, whereupon the owner pointed to Geoff's house opposite. We thanked her and went and banged on the door, which a bleary eyed Geoff eventually opened, to be told he had a son. After swallowing some black coffee, he dressed and went to find his wife. It being St George's Day, the baby was christened George.

We hosted Ford dealer visitors from around the world at Boreham House, and a group that particularly sticks in my mind is one from Norway. They would tell the barman to take a pint glass and add one measure of each of the optics behind the bar, and then fill it with bitter. They explained that alcohol was so highly taxed in Norway they couldn't afford to drink. After a couple of these cocktails they were completely drunk and usually slept in a chair in the hall.

Even more memorable was a group of three Canadian zone managers who came to Boreham to learn about the Dexta and Major. They were great guys and lots of fun, and were determined to make their stay something to be remembered.

Their plan was to disassemble a Dexta in the car park of Boreham House and reassemble it in the bar, and as I was on duty that night I decided it was easier to help them than try to stop them! The bar at Boreham was in the cellars, to which the only access was down a narrow staircase or through the old coal hole. The Canadians chose the coal hole through which to transfer the tractor parts, and proceeded to remove the tractor's wheels, engine, gearbox and rear transmission and, using rollers, then tease them through the hole and down the passage.

They were all strong chaps, and I looked on in disbelief as one of the biggest, Steve Nessner, actually picked up the engine and walked through to the bar with it. However, the stunt almost came to a grinding halt when the 10 × 28 rear wheels wouldn't fit through the coal hole. By letting the air out of the tyres and binding them with rope, though, they just fitted down the back stairs. By 6am the engine, gearbox and rear transmission were joined and the engine was running. We all went to bed for two hours before the tractor was discovered, and the Canadians left that day, leaving Boreham staff the job of removing it!

***

To maximise exposure of its tractors, particularly the Dexta and the Power Major, Ford planned a series of five farming fairs during 1958, the most notable of these being at Ashton-in-Makerfield,

Lancashire. It was an event that created such memories that all these years later I still have no trouble recalling the occasion.

The idea was to make it a show for the family: for the father there would be the chance to examine the latest tractors; for the mother the opportunity to see a top-of-the-range fashion show; and for the children we had a playground. It really was all the fun of the fair.

The site was a 100-acre block of land reclaimed after being an open-cast coal mine, and the fair lasted three days, with every trade you could think of being represented. Such a show would never work in this day and age, but five dealers joined forces and invited thousands of farmers to attend. It was probably the biggest demonstration of tractors ever staged in this country.

We had four Thames Traders to carry Ford promotion exhibits, including a radio-controlled Dexta, fitted with a single furrow plough that could be lifted out and lowered into work. A lot of tractor drivers were worried that a man would be able to sit in his office and run ten tractors at a time, but to alleviate fears it was pointed out you needed to be highly qualified to operate one and you still needed a man to change the plough shares.

People came from miles to see around fifty Ford tractors working on the site. We also had County Full Tracks, four-wheel drives from Roadless and Muir Hill, loaders, ditchers, hedge cutters, a full range of equipment from Ransomes and other implement makers, and even a display from the winch makers Boughton and Cooke, using their devices to remove tree trunks.

In order to keep the crowd together we demonstrated each tractor in turn and gave a commentary to match. This was done every hour. The fashion show for the ladies was also hourly, with a cookery demonstration in between. Probably the biggest attraction, though, turned out to be two very pretty girls trying to sell Stork margarine. They happened to be staying in our hotel, and I remember one particular dealer falling madly in love with one of them.

On the last day of the show we had one of the worst rainstorms I have ever seen. By lunchtime the reclaimed coal mine was 6in deep in black mud, and ladies were trapped on stands unable to get out of the fair in the shoes they wore. The whole site was in chaos, and all the cars had to be towed out. By 9pm most of the Ford people had moved into the dealers' caravans, and I can recall being in the one owned by Tom Quick, of Quicks of Manchester, while Peter Kirby, from Kirby's Farm Services in Wrexham, was next door. Gordon Guthrie, who was our tractor manager, was also stranded with us, together with his wife. At 10 o'clock we decided to try and get out, and Tom offered to carry Gordon's wife towards dry land. He took two steps and fell flat on his face, tipping her into the mud, her white fur coat turning as black as the look on her face ...

We decided to go to the hotel in the Thames Traders, which were parked nearby, but first had to tow the trucks to the roadside. I got a Dexta, and tied a rope between the two vehicles, failing to notice the rope was a hundred yards long, and only realising once I'd driven to the top of the site and the truck still hadn't moved. Eventually we escaped and got back to the hotel, but once in the bar we realised we'd lost Peter Kirby. He turned up much later, having driven fifteen miles on an unlicensed tractor with no cab, and more than likely over the drink-drive limit. He was so cold and wet he could barely speak. I now look back on it as one awful day.

***

It wasn't long before I got the job I had always wanted: southern area manager, covering Kent, Sussex, Surrey, Hampshire and Wiltshire. This was very much blue countryside, with at least a 40 per cent share of the market going to Ford. We had a wonderful set of dealers, including Haynes, Invicta and Stormont, who were all Ford triple franchise businesses selling cars, trucks and tractors. The Kent Ford dealers (KFD) group even built their own modified orchard tractor, a low-profile Major with narrow half shafts and small wheels but with lots of power for fruit spraying. The launch of the Dexta, though, reduced the demand for such conversions.

Many of the triple franchise dealers had started their businesses selling Model T cars, before later taking on the tractor franchise during the Second World War. Other names included Crimble of Staines, Rowe of Chichester and Gilbert Rice of Horsham. Once into the big farming areas in Hampshire and Wiltshire, we had tractor-only dealers – Watson and Haig, Brewer and Co, and TH White, one of our biggest tractor dealers. Tommy Lowe of Sussex Tractors was the Massey Ferguson dealer for the county, based in Uckfield, but during the 1960s he had been told by MF that he must sell its combines as well as its tractors. This he refused to do, and Sussex Tractors became a Ford dealer. Later on he employed Paddy Campbell, who had been the parts manager for Gilbert Rice at Horsham, but became a great salesman.

It was from this period I remember one of the strangest mechanical problems I ever came across, in terms of the cause of the problem. I was at a dealership in which there was a Major in the workshops. It was producing next to no power, and I had helped to try every idea we could think of to solve the problem – even changing the injectors, which seemed to be in good order, and ensuring the injection pump was correctly calibrated, which it appeared to be.

It was then that we paid a little closer attention to the air cleaner. After removing the well that held the oil, we then found to our surprise that it was full of paper wrappers. Upon consulting the owner of the tractor, we discovered that his driver had an addiction to toffees, and was in the habit of amusing himself by flicking the wrappers at the air cleaner – which had lost its 'tin hat' cover –

and scoring a 'goal' when one entered the pipe! We cleaned out the air cleaner and bowl filter, fitted a new inlet pipe and tin hat – and the tractor ran perfectly thereafter.

Sometime during the 1960s I took a party of dealer salesmen from this zone on a two-day trip to Hamburg, awarded for meeting sales targets. One night we went to a beer fest, with an oompah band playing in a huge hall, and lots of waitresses in national dress carrying multiple jugs of beer. This led to a very jolly evening, with lots of beer and singing. At the next table was a group of three very friendly and very drunk German girls, who spoke very little English. Somehow when it was time to go, I ended up in their VW Beetle.

We hadn't gone far when we realised we had gone the wrong way down a one-way street. It was unfortunate that, at this point, a policeman appeared out of nowhere and stopped the car. The driver was asked to get out, and I did the same. She was asked to blow in a bag, found to be way over the limit, and was about to be arrested. The policeman then turned to me and asked me something in German, whereupon I told him I was from England. His tone immediately softened, and in near-perfect English has asked me what part of England I was from. It turned out he had spent two years in Ilford, just 20 miles from Basildon.

We had a long conversation and, somewhat unbelievably, he suggested he would turn a blind eye if I was to take the girls to a nearby café and feed them black coffee for a couple of hours. This was no hardship for me! When it was time to go, though, I had no idea where I was. Fortunately, my room key had the hotel's address on it and I made it back safely.

4

Dealers and Customers

During my time as southern zone manager we bought a new house in the area, in Midhurst, Sussex, and I called it Dexta's, after my favourite tractor. The meaning of the name certainly provoked a lot of questions. My son was born in the house, arriving at 3am one day when Bill Batty, head of Ford's UK tractor operations, had called a meeting at 8am the same day at the Ilford offices. One of our zone managers was late for the meeting and was told he shouldn't have bothered to come at all, but I was on time having had one hour's sleep! It was the late 1950s and in those days it didn't matter – we wanted the job and did what we were told, particularly if Mr Batty was doing the telling ...

Bill Batty was very much the boss, and was also a director of Ford of Britain. It was always 'Mr Batty' to his face but 'Bill Batty' when talking to a colleague. Top management in those days needed to be tough and respected, and he was both of these things. If we went out to dinner after a day at the Smithfield Show he was one of the boys and treated us all as friends, but the next day it was 'Mr Batty' again, and everyone had to be on the stand at 8am. His pet hate was suede shoes, which were very fashionable at the time.

In 1958 we launched the Power Major, a small increase in engine power over the Major, justifying the new name. However, the engineering department had failed to inform us that, for some reason they had decided to incorporate a small reduction in the ratio of the crown wheel and pinion, meaning a slight reduction in speed in each gear. A week after the launch I received a phone call from a baling contractor in Kent saying that this minor speed change meant that he lost 200 bales in a day – most of his profit. We solved the problem by replacing the old crown wheel and pinion, and latterly offered customers a choice of ratios, before later moving wholly to the faster ratio.

At the same time, we were also looking at ways of selling more of our smallest tractor, especially as loyalty to MF 35s by their owners was very strong. Each zone manager was allocated a new Dexta and told to place it on a farm using Massey Ferguson 35s. On the Hog's Back in Surrey there was a farmer, Bob Hewitt, who grew about 1,000 acres of lettuces and radishes for Covent Garden and had a fleet of nine MF 35s. I chose him as the ideal customer to have the new Dexta. He liked the idea, and soon, with the help of the local dealer, Hyde Abbey, we had the tractor installed among the fleet of Fergusons.

I used to call in every week to make sure he and his staff were happy with the tractor and report back on what it had been doing and how many hours it had worked. On one visit, Mr Hewitt himself came to meet me, and took me for a drive in his new car, an AC Cobra. I climbed into the passenger seat and very soon we were doing 100mph. I told him I was very impressed but he suggested I hang on as we hadn't yet reached the correct operating temperature. We watched the gauges and we took off for the second time. As I looked down at the speedo we were doing 140mph. It really was an amazing experience – I had never even achieved 100mph in those days. On the return trip, I was further amazed when he suggested we swap seats and I drive back to the farm. He told me to give it some welly, and every time I touched the accelerator the car seemed to launch itself up on to the crown of the road. We did get back to the farm in one piece but I had already made up my mind that it had scared me enough not to do it again.

The Dexta had been on the farm for three months, with the option to buy it at a special price. After a lot of bantering with the dealer, Bob Hewitt bought the demonstrator and traded in all of his nine Fergusons against Dextas, the biggest single order for that model I ever took.

I hadn't been on the territory long when the man looking after the West Country was moved and not replaced, and I was given Dorset, Somerset, Devon and Cornwall to add to my southern area. These counties were Massey Ferguson strongholds, and instead of the 40 per cent market share on my own patch, Ford tractors accounted for only around 10 per cent of the market here.

Life was good fun, with wonderful dealers such as TH White in Wiltshire and the surrounding counties, Watson and Haig at Andover, Hants, and PA Turney in Oxfordshire, previously a Nuffield/Leyland dealer. I had signed Turneys relatively by chance. One day while out on territory in Oxfordshire, I happened to drive past a field with a big 'Tractor Demonstration Here' sign. Curious, I drove in and happened upon a Leyland demo day being held by Turneys.

Within a few minutes, I had introduced myself to Peter Turney and was sitting in his Land Rover. I learned that things were not going well regarding Leyland sales, but his other franchises, such as combines and forage equipment from New Holland, with whom we worked closely, were doing well. I said it would make me very happy if he were to become a Ford dealer. About two months later we celebrated his doing just that with a wonderful Chateau Talbot; I remember it so clearly as it was the best wine I had ever tasted.

The Major, Dexta and the Ferguson 35 and 65 made up nearly 80 per cent of the UK market during the 1960s, with International, Nuffield and David Brown representing most of the remainder – John Deere was relatively unheard of at that time. Our dealer strength was in the predominantly arable east of England, while Massey Ferguson dominated the West Country. Dealership strength went in line with tractor sales – had Ernest Doe been born in Somerset he would certainly have been a Massey

Ferguson dealer, but fortunately for Ford he was born in Essex!

The Major gradually gained more power through new guises as the Power Major and Super Major. I remember receiving a phone call from Colchester Tractors, part of the Doe organisation, asking if we would demonstrate the Power Major against the MF 165. They hired the MF with a tractor driver and it was up to us to compete in the field, side-by-side. It was quite a clever move on the part of Colchester Tractors to drive home to us the problems caused by our lack of draft control to give better wheel grip on wet ground.

Had they picked a much wetter field the MF would have shown up the Ford, but field conditions were good and at the end of the day, with tempers a little fraught, there was very little to choose between the two tractors. In all fairness we didn't need to be told our hydraulics were out of date, but our engineers did ...

By 1960, though, we had the Super Major with its all-new draft control, differential lock and disc brakes. This was a thrilling time for all of us who were regularly called on to demonstrate and help out farmers with field work problems. The hydraulics were excellent, and we could happily take on MF in the field without any worries.

When asked by farmers to view a machine with an operating issue, I liked to impress by wearing a dark suit and polished black shoes, giving the impression I had no idea how to drive a tractor. If both wheels were spinning and the work output was nil, I would at the last moment put on boots if it was very wet, but if it was dry I wouldn't bother. I'd ask the driver to let me take his seat, and would almost always find the draft control lever in position control.

After flicking the lever to draft and dropping the engine speed to 800–1,000rpm, with luck I would be motoring up and down the field within a minute. The suit and shiny shoes were my gimmicks, and while the farmer was delighted, I could walk away having left a good impression for a man who might have looked initially as if he had little experience of operating a tractor in the field!

Tractors of 30–60hp made up the bulk of those sold during the early 1960s, and although I'm sure that if any UK tractor suppliers had been able to break the 100hp barrier at this time the tractors would have sold, engineers were cautious of the unknown effects of such power on long-term reliability. There were plenty of tractor drivers on the farms, which were still small on average, and field sizes had not yet grown through the removal of hedges and ditches, which came later. I can remember several farmers who purchased forty or fifty tractors at a time, such as Banks of Spalding and East Anglian Real Estate. One farmer from Wiltshire, a Mr Wookey, who was a great friend of Peter Scott's, brought all twenty of his tractor drivers to Dagenham for the day to see their tractors being built on the line, each being photographed with what would soon be his own machine.

5

Launch of the 6X Line

In early 1964, it was suggested I needed more of a challenge, and I was appointed area manager for south Wales. We moved to Winchcombe, near Cheltenham, and if I thought Devon and Cornwall was difficult, at least they spoke English! By this time we had two very small children and my wife had a hard time bringing them up on her own.

It was also a challenge to boost dealer sales numbers in my new area. While across the Kent to Wiltshire region Ford had a market share of 40 per cent, in south Wales it was half that. However, my stay was to be a short one, and just as we were settling into another house, I received a phone call to say that I was to be the new manager of my old haunt, the Ford tractor training centre, Boreham House.

It was a big promotion – one day I was a Grade 8 zone manager, and the next I'd jumped up the pay scale to Grade 11, at 29 years old. However, I was told that I couldn't use the Boreham car. Subsequently I bought a second-hand MGB, parked it outside the front door of Boreham House and refused to move it. It wasn't long before I had the company car for my own use!

This was a big time for Ford for two key reasons. The first was the move of tractor production from Dagenham to a new, purpose-built factory at Basildon, just a few miles away. The second, which would happen at the same time, was the introduction of the new 6X tractor range – the 2000, 3000, 4000 and 5000, later to be known as the 'pre-Force' line.

Over a period of two months, all our dealers and their salesmen were coming to Boreham House for a three-day sales course. Initially, though, I was still living near Cheltenham, and had only been in the house for five months. My wife had now got two children under five years old who had to be pushed up and down a very steep hill to the nearest shop, while I had a three-hour drive to Boreham.

With a full six-day week of sales courses over nearly three months I was trying to buy a house near to Boreham, but my wife had no way of helping me as she was still at our existing house. However, I eventually found somewhere in Chelmsford we liked the look of, with the advantage of it having building permission for a house in the garden. I had a very bright solicitor who realised there was no access to the plot, and that resulted in the price dropping by £2,000, whereupon I bought it. With its proximity to London, Essex has never been the cheapest place to buy a house, and it was notable that the area managers from the north or west of the UK showed little enthusiasm for promotion within Ford, as it would invariably mean a move to be closer to the factory, where they would get much less property for their money.

We were due to move house from Winchcombe to Chelmsford on a date that fell after the last day of a training course. After a very late night in the bar I left Boreham at 5am, but just as I reached the dual carriageway leading to Oxford, a man in a Mini crashed into the side of my car. Fortunately, no one was hurt, though, and I was able to hire a car from the local dealer. I was two hours late meeting my wife on moving day, which was very worrying for her at the time, but ultimately all was OK.

I went to Boreham House just six months before the launch of the 6X Ford 2000–5000 tractors and the move of production from Dagenham to the brand new Basildon factory. In 1964, when we moved from Dagenham, Massey Ferguson was just ahead on national sales, being particularly strong in the dairy areas of the west country, Wales, the north-west and the west of Scotland.

We were very close in second place, our main business coming from the arable areas in the south, East Anglia, Lincolnshire, Yorkshire and the east of Scotland. Life was good at Dagenham, and in our last year with the Major and Dexta we made $20 million profit for the parent company.

The new Ford 6X range was a joint venture between Ford HQ in Detroit and the team at Dagenham. I had no experience of Detroit-built Ford tractors, and nor did I know anything about American farming, but as time went by, I came to realise that the American farming scene was totally different to Europe. For example, most field-working equipment in Europe was mounted on the tractor's three-point linkage. In America, everything was towed on the drawbar and raised/lowered using the tractor's hydraulic valves.

Pricing of the four new tractors was very important, and it was crucial to keep below £1,000 to be competitive – the 5000 listed at £998, which included a set of rear wheel weights. At that time, it was unheard of to price a tractor over £1,000 or a combine over £3,000.

There were no Ford cabs at the launch, but the new machines could be equipped with British-built outside-sourced units from the likes of Duncan and Lambourn. They at least kept the rain off, but they were still cold and tended to exaggerate the noise from the rear transmission.

My first job at Boreham was to teach the whole European dealer organisation about the new 6X 'Thousand Series', a massive year-long programme involving two sales and two service courses a week. It was the custom at Boreham to have a question and answer session on a Thursday evening with visiting senior managers from Basildon, and the first question, invariably, was why we were using an eight-speed gearbox in which two gears matched, making it in essence a seven-speed transmission. There was no answer to this that would satisfy the salesmen.

The other comment that came from tractor drivers using bigger and bigger spanners on heavy tackle was that the tool box was far too small. It was deemed big enough to keep your cigarettes in provided you weren't a chain smoker – although the less polite would say that it was big enough to keep your condoms in providing you weren't a sex maniac!

But while these were seemed relatively small gripes, they were compounded by the fact the launch was an all-time disaster. The Fordson Major was tried and tested and had roughly 40 per cent of the market. Then came the day we had to move out of Dagenham into the brand new factory Ford had built for tractor production on a 100-acre site in Basildon. The biggest problem here wasn't the facilities, though – it was the labour. Dagenham people weren't especially keen to move to the Basildon plant, and sadly many of the new staff taken on were untrained, unable to recognise a wrong-fitting gasket or a crossed thread. This did nothing to help us out of the trouble we were in.

The big day came to begin production of a new range of tractors at the new factory, but very quickly we were desperately short of parts, with incomplete tractors being stored anywhere there was space – there must have been 100 in the gardens at Boreham House, and many others at various industrial sites in Basildon, all short of vital parts, unable to be sold. Unfortunately, it was our most loyal customers who were suffering from our problems. They were desperate times and there seemed to be no answers.

I well remember going to a meeting to discuss the failure of the balancer gears fitted to the 5000, of which we had sold 12,000 across Europe. These gears ran off the crankshaft, tooth-to-tooth. One day there was one failure, and by the end of the week we were having five a day. It was discovered that the teeth had been over-hardened, causing them to become brittle and fail at about 800 hours. If you were lucky, the teeth broke and lay in the sump; if you were unlucky, they broke a hole in the side of the cylinder block. After a hastily-convened meeting on the issue following a number of reports coming in from customers and dealers, it was decided to spend $1 million on a campaign change rather than risk the possible cost if we did nothing. Fortunately, it was not a big job to drop the sump, undo the securing bolts and fit a new crankshaft.

The next and much more challenging issue, though, was porous blocks, as one after another farmers were finding problems with water leaks in the cylinder bores. It took a very long time to resolve this issue, with some being repaired by boring out the existing bores and fitting new dry-liners. The problem was eventually solved by fitting a water filter to the engine, similar to those already used on other tractors on the UK market. But the term 'porous Ford' became legendary.

Although it had not been without its issues – mainly due to driver education on operation – the clutchless Select-O-Speed transmission remained available for the revised Ford Force 6Y tractors launched in 1968. During dealer introductions at Boreham House, my colleague, Martin Brown, would drive a Select-O-Speed tractor at full throttle in top gear into the yard before groups of dealer salesmen, before pushing the lever into reverse, whereupon the wheels would spin and the tractor would proceed to disappear out of the yard the same way from which it came, invariably to a round of applause.

It was a wonderful gearbox, but reliability was a major issue. For whatever reason it seemed to work best in the 5000, but in the 4000, in which the box had to be turned upside down for fitment, it was particularly unreliable. The situation in the field was so bad that we were offering the dealers eight-speed standard gearboxes at a special price to replace the Select-O-Speed ones.

But the dealers had grumbles with the standard transmission too. At the end of the three-day dealer course, we returned with a group of dealer staff to a meeting room where they were invited to make their comments, which I then wrote on a two-sided blackboard. The first criticism was our choice to call it an 8 × 2 gearbox when in fact it was a 7 × 2, with two overlapping speeds. I recall someone pointing out that it was better than the old 6 × 2 in the Major, but that was not a lot of consolation to the salesmen.

The complaints continued. There were no flat deck operator environments. There was a shortage of pick-up hitches – in fact there were shortages of everything due to problems with component supplies at Basildon.

When we had a full blackboard of complaints we used to reverse the board and there would be an identical list of problems on the other side. That was to prove we knew what was wrong and that we were doing all we could to correct the issues.

Sales were suffering from a number of problems, and our market share was sliding from 40 per cent towards 20 per cent. The telephone never stopped ringing with irate customers wanting to get rid of their unreliable tractors. One such call came from John Alvis, a charming man who farmed near Redhill in Somerset, and was a very big cheesemaker and pig farmer, feeding his pigs on the whey. He had owned five Nuffield tractors, which he had traded in for five Ford 5000s, but the tractors had had numerous problems, and he declared he wanted the Nuffields back. I recall driving across to see him in my Lotus Cortina. His dealer, Leslie Jewell, and I met John in his office, whereupon he produced a bottle of whisky, unscrewed the top and dropped it into the bin.

'No one is going to say I'm not hospitable, even if we have some problems to talk through,' he said as he poured the bottle into three tumblers.

John never raised his voice, and after we had finished lunch with him we sat down and had a deal for five new 5000s, at £980 apiece including a set of rear wheel weights. John Alvis was a happy man and so was I, particularly as I got home without seeing a single police car – the day of our meeting happened to be the day drink-driving laws were introduced! John and his two sons became very good friends of mine, and many years later even sent two blocks of cheddar for my golden wedding anniversary celebrations held at Boreham House.

***

To add to the headaches caused by tractor reliability issues, I was suddenly faced with the departure of our housekeeper at Boreham, who had decided to get married and leave her position. It was a vital role, the equivalent to the manager of a hotel for 30 people, and I had absolutely no knowledge of how to find a new one. The big advantage was that a cottage went with the job, and when we advertised we had about 150 applicants. We short-listed about 25 people, and I interviewed all of them, but not one applicant was vaguely suitable, and I began to despair. I was on the verge of giving up when I received a late application from a Mrs Pamela Ives, who had been running a very smart pub on the River Thames but was getting divorced and looking for a new job in a different area.

Instead of dotting her 'i's she put a tiny circle over the top of them, which was another reason her application made an impression on me, and I invited her for an interview the next day. A smartly-dressed visitor arrived promptly at the appointed time asking for Mr Pearson, and I told her she had found him, to which her response was that she had been expecting someone much older! But Mrs Ives had every possible qualification to do the job, and was promptly offered the position as our new housekeeper, which she did excellently for many years.

Boreham House was not only a wonderful place to work but it was a magical place to entertain farmers and a great asset to the company. For a start, it was the only Ford location in England where you could have a gin and tonic or a glass of wine! Most farmers knew where Boreham House was, and as soon as they came through the gates and saw the house and the lake in front of it, half of their problems went away. Before the arrival of customer visitors, I would ring Mrs Ives and arrange for lunch in the VIP room. I always kept to the same menu: sirloin of beef followed by apple pie.

On their arrival, I would take them for a tour of the classrooms, telling my audience how serious we were about getting the job right. We then returned to the VIP room for a couple of drinks followed by lunch and wine. At the end of lunch I would suggest we talked about any problems and issues. However, by this time they were almost apologising for even suggesting they had any! We waved goodbye on the front steps to happy customers.

***

As a keen tractor driver myself, I always yearned for more horse-power. During the early 1970s, I saw an advertisement in the Farmers Weekly for a Northrop turbocharger kit to fit the Ford 5000, at a cost of £120. We had some money from bar profits at Boreham House, so I invested in a kit, and an hour after its arrival we had created the very first 7000. It gave the tractor an extra 19hp, taking it to 94hp, and was unbelievably quiet in comparison. The 5000 to which we fitted the kit was a Select-O-Speed tractor, to which we also fitted a County crown wheel and pinion that gave it an additional 10–15 per cent more speed throughout the range. In order to have it tested, I sold it to my very good friend in Hampshire, Denys Cope, son-in-law of well-known Hampshire farmer Rex Paterson, and it went on to record more than 10,000 hours. Persuading the engineers to adopt the idea, though, took a lot longer.

I immediately wrote to our chief engineer, suggesting that he get on the bandwagon and introduce the turbocharger. The extra horsepower was wonderful when implements, particularly PTO-driven ones, were getting bigger and requiring more and more power. The added bonus was the decreased noise level in the cab. The turbo had a low whistle and was very popular with tractor drivers spending all day in the seat. Despite it taking a while to convince our engineering department of the concept's merits, once it was eventually launched in 1971 the 7000 became a strong seller in no time and made good money for the company. It was also the first Ford tractor fitted with the Load Monitor lower link sensing system, an excellent Ford innovation.

Having been at Boreham House for the worst two years in Ford's tractor history, owing chiefly to those 6X reliability problems, I moved back to Basildon to become southern area sales manager, overseeing five zone managers and the dealers they looked after. It was a job I really wanted, but not at this particular point. Although Ford's share of the market was around 20 per cent, I seemed to spend all my time visiting irate customers. Over the next two years, though, the engineers gradually improved our tractors' reliability, and the balance of power gradually shifted, with Ford regaining its market leadership over Massey Ferguson across the southern area, making me a happy man.

6

Good Times, Bad Times

By the early 1970s, agriculture was in a depression, with knock-on effects for the tractor and machinery markets, and farmers were saying they would never again be able to afford a new tractor. But as the middle of the decade came round, after the drought of 1976 wheat was £140 a ton, and the unbelievable happened. Tractors were on allocation to the dealers and we had begging phone calls from farmers desperate to spend money and avoid paying tax.

I had never believed this could happen and I think it was worse than having to fight for every sale. The problem was that we couldn't build tractors fast enough, and couldn't source sufficient key components such as tyres. Anticipating a full recession, the tyre manufacturers had scaled back production, and were neither able nor willing to build new moulds to increase production. This situation lasted for about 18 months before the market stabilised and business was back to normal.

There were some lighter moments during the dark times for Ford and agriculture in the 1970s. Among those in which we were involved was the annual flower parade in the Lincolnshire town of Spalding, known as the bulb-growing capital of the UK, which featured highly elaborate and beautiful tulip-decorated floats either pulled by or mounted on tractors. The floats always had a theme and the designs were prepared early in the year, with the base of the float being blacksmith-made and then covered in wire mesh. Literally thousands of tulip heads were then attached to the netting with hair grips.

We liked to support the event, and prior to safety cabs being a legal requirement it was no problem supplying Ford tractors well in advance of the parade, so they could be made ready for the tulip heads to be pinned on. However, with the advent of compulsory cabs we had to borrow tractors destined for Japan to use for the festival, as this market did not have safety cab regulations.

I always received an invitation to partake in the event lunch and then watch the parade from the grandstand. The floats were quite unbelievable and the parade would go around Spalding before ending up in a field, where they could be admired by the thousands of people who came to see the show. Having seen the parade twice I was then able to pass the invitation on to other members of the company. Sadly, today the parade has ceased to exist, as companies are no longer able to afford the sponsorship of the floats, but it has recently been replaced by the Springfields Festival.

***

Some areas of the UK had some particularly large landowners, and not all of them were British. East Anglian Real Property was actually a Dutch business that came to Norfolk to grow sugar beet and barge the roots back to Holland to process them into sugar, in the days before the county's Cantley factory was built.

One day at the Royal Show, Rob den England from East Anglian Real Property came on to the Ford stand, where he met Alistair MacGuillivray, the northern area manager, who asked if he could help. When he was asked if he could discuss the possibility of placing an order for 47 Select-O-Speed 5000 tractors, Alistair quickly came to find me!

Larger farms required not just more tractors, but larger ones too, and farmers all over Europe were desperate for more and more power. The Americans had already seen the need for this

42

and came up with the 8000 and 9000, later succeeded by the 8600 and 9600. Big tractors with six-cylinder engines, they were ideal for trailed implements, but they had no four-wheel drive, which was becoming a must on 100hp-plus tractors in Europe, with its fully mounted equipment and heavy wet land to be cultivated.

These two big tractors worked well in dry land conditions after harvest when the straw had been burned, though, which was common practice in those days. But without air conditioning it was unbearably hot in the cabs, which were fitted on this side of the Atlantic and supplied by Swedish firm Hara. There was a choice of opening the doors or the roof hatch, but that meant being choked to death by the soot from the burned straw!

We got what we wanted in the end, a four-wheel drive system built for us by the German specialist axle company ZP (Zeppelin Passau). It was essential to ensure, though, that the lever that provided the drive to the front axle was released – leaving it on caused it to overheat the discs and the drum and resulted in a serious oil leak. By the time we got the TW range of high-hp machines in 1979, there was a more European focus on big tractor production, resulting from its move to our plant in Antwerp, Belgium. Latterly, this factory switched to manufacturing components, and from 1994 big tractor production went back to North America, or Canada to be precise, being switched to the former Versatile factory that we acquired in 1987. But by this point there was a much more worldwide focus on the needs of farmers globally.

***

Returning to the mid-1970s, in late 1975 we introduced the 7A1 range, or '600' series', to succeed the 6Y tractors. Originally these tractors had been due to have a new cab design, but this was delayed and the new machines were launched with the old Fieco safety cabs. The 2000, 3000, 4000, 5000 and 7000 were replaced by the 2600, 3600, 4600, 6600 and 7600, and additional 4100 and 5600 models were added to fill horsepower gaps. Engine and hydraulic upgrades were the key improvements.

The following year, though, government legislation meant quiet or 'Q' cabs, which shielded the driver from noise above 90 decibels, became mandatory. Telford-based GKN Sankey was responsible for the design of our Q cab, which would be fitted to every model in the main Ford tractor range. I remember seeing the first prototype and stepping into the cab, sitting down in the seat and realising something was wrong. The seat was resting on the rear transmission to avoid the operator hitting his head on the roof. Eventually the roof was raised and the problem solved.

Next, though, came implement compatibility tests. I felt it was a brilliant cab, quite the best on the market, but there was still an issue or two. Reversible ploughing was becoming commonplace, and the typical Ransomes plough being used by farmers at the time had a hand-operated turnover. The first Q cab prototype had a cross beam at the back of the seat, and the handle mounted on this fouled the cab when the plough was raised. Our engineers were struggling to meet the compulsory 1976 deadline for the introduction of Q cabs, but eventually the beam was removed, we had a cab that passed all the tests, and we were in business.

I created the presentation we were to make to our dealers on the Q cab, using a sound film we'd had made to show the difference in noise over the old safety cab, and I was very proud of it. On one occasion we were doing a practice run-through when, after about ten minutes, Laurie Chivers, our sales director, asked if I had a script for the presentation. I told him I hadn't and preferred to do things from memory while working from the slides. Asking what the company would do if I dropped dead, he said he would prefer it if I wrote a script! This I duly did, placing it on the podium before me, but I never once used it. Laurie never mentioned this to me again.

The presentation began with a Boeing 747 on take-off at Heathrow, producing a sound level of at 145 decibels, and then worked through a series of road drills, London buses and other noisy machinery until it reached the 85 decibel level in the new cab.

Our main competition at the time, Massey Ferguson, had launched a new cab with a single left-hand door, which wasn't proving very popular. To wind the dealers up a little, I put up a slide showing the left side of the Ford cab and said:

'With this new cab, there is only one door ...'

General murmuring and grumbling from the dealers ensued, before I continued:

'... on the left-hand side, and one door on the right!'

There was an enthusiastic response, with lots of cheers. Double-decker lorry-loads of the Q cabs were a common sight on the M1 as they made their way down from Sankey's Telford factory to Basildon. With its two doors, very low sound levels, a heater for the winter and optional air conditioning for the summer, tractor drivers were delighted with the luxury of the new Q cab, variants of which were ultimately fitted to every Ford tractor from the 2600 to the TW-30 introduced later that decade, in 1979.

Although we had developed the Q cab for the lower-horsepower tractors, this was a 'straddle' design, with the transmission tunnel between the driver's feet. Czechoslovakian tractor maker Zetor had introduced a flat deck Q cab, which had proved very popular with tractor drivers, and it was clear we needed a similar layout. I recall meeting the Duke of Gloucester one year at the Royal Smithfield Show, where he told me his tractor drivers had a fleet of Zetors with which they were very pleased. When we came up against farmers who included their tractor drivers in the decision over which tractor make to purchase, the driver often said he favoured a Zetor over a Ford because of the cab comfort, and this influenced a number of buying decisions, with little regard for resale value or parts availability.

Ford engineers in America had been working on a flat floor design for the 6700 and 7700 tractors that were planned for Basildon production in 1976 as super-luxury 'rowcrop' versions of the 6600 and 7600, with 16 × 4 Dual Power transmissions in place of the standard 8 × 2 unit. The tractors were to be sold alongside the 8700 and 9700 replacements for the 8600 and 9600.

My colleague, John Power, and I were invited to go and see the cab development at the Ford engineering centre in Detroit. Transmission operation was based on an American column-shift gear change design, some years before a similar concept was introduced on the Basildon-built four-cylinder Series 10 tractors. While this latter transmission used a lever mounted on the left of the steering column, this earlier design engineered by the Americans for the flat deck cab had a shift mechanism on the top of a column mounted between the driver's legs.

John and I considered it unbelievable that anything so ridiculous could have been designed, and had the job of telling the vice president of tractors that the design was unsaleable. As you can imagine, this didn't go down very well, but management did listen to what we said. As a result, the tractors were ultimately fitted with much better right side-mounted gear and range levers. We launched the 6700 and 7700 in August 1976 at a university venue in Finland. At 78hp, the 6700 was a bit gutless for a big tractor, and greater numbers were sold of the 97hp 7700. At the same event, the new Antwerp-built 128hp 8700 and 153hp 9700 replacements for the 8600 and 9600, which were available with Ford's first 4wd option, were also unveiled.

It was about this time that a Belgian Ford dealer invented a Ford tractor that started on diesel and then ran on gas produced by a side-mounted boiler that burned coconut husks. It was only viable if you had palm trees on the farm! However, a slightly more successful invention of the same dealer, EVA, was the 8100. This six-cylinder 115hp tractor, which was available only in 2wd, unlike the 87/9700, was first sold in France, where the concept was popular as a machine for long-distance, high-speed haulage. It was based on a 6700 back-end, with the engine supported on lengths of steel plate.

The design seemed a good one at the time, and the 8100 was launched in the UK in 1978, where we used the extra power available at the PTO to power equipment such as forage harvesters – the tractor was first shown at the National Grassland Demonstration. However, the tractor didn't prove a good fit with the UK market, and the 8100 was later replaced in 1980 with the County-built 8200, which had standard 4wd. Meanwhile, in 1979 the 8700 and 9700 were replaced by the new Antwerp-built TW-10 and TW-20, with a new 188hp TW-30 at the top of the line.

***

One of my greatest local farmer friends was Tony Speakman, from Sandon in Essex. I first met him by chance when I gave his son, Kit, a lift home from a children's party, after which Tony invited me in for a drink. He was always the most hospitable of people, and hours later I said goodbye and got into my three-litre fuel injection Capri, a very skittish car at the best of times. On a single track road, I missed the corner and ended up sitting in a ploughed field. After walking back to Tony's house, he came to help me, but just as he was about to pull me out a police car with all lights flashing appeared on the scene. An officer came to my window and suggested it would go better in reverse. Once I was back on the road, he advised me to be careful on the main road, as he was with the dog police, and others might take a more dim view of my activities!

This was the start of a great friendship between Tony and I. Sometime afterwards, we visited a neighbouring farmer, Fred Tiley of Little Claydons Farm. He was a dairy farmer, and adjacent to his dairy unit was the biggest muck heap I had ever seen. Fred offered Tony the dung for his land for free if he would remove it, and Tony drew me into the challenge. I organised three 30-ton tippers, borrowed a Ford A66 loading shovel with a three cubic yard bucket, and work commenced. It took some 200 loads at 30 tons a load – 6,000 tons of dung had been carted five miles to be spread.

We had performed marathon operations before at Ford, as publicity events to bolster our reputation for reliability. These included a muck spreading marathon on the dairy unit at Lord Rayleigh's Farms, not far from Basildon, a 24-hour ploughing exercise, and a marathon forage-harvesting stint carried out on Rex Paterson's farm.

However, this spreading was to be a one man job and a personal challenge, using two Howard spreaders, two Ford 7610s and the A66 loader. We had launched the Series 10 tractors in 1981, replacing the 2600–8200 with the 2610–8210, with one of the main – and ill-fated – upgrades being the little-liked column gearshift for the 7610 and below, designed to create a semi-flat floor (the transmission tunnel remained).

I started early in the morning, filling both spreaders before spreading both loads and repeating the process, and by 9 o'clock that night the job was done. Several weeks later Tony invited my wife and me to go to Chelmsford theatre, where we saw English folk group The Yetties, who very appropriately sang 'Fling it here, Fling it there' just for me, before he presented me with a recording of the song.

Tony was an asthma sufferer, and on a very hot and humid night in 1982 he died. It was a sad time for me, compounded by the loss of sight in my left eye. One Monday evening that year,

Geoff Tiplady had asked me to attend a farmers meeting with the Huelva group of Cornish farmers, who were named after the people who announced the arrival of the pilchard shoals off the coast. I travelled down by train with my car, which you could do in those days, so I could visit dealers on my way home. I got into my car at Truro station and thought the windscreen looked very dirty, but assumed it was coated with dust from the train.

Having cleaned the screen. I got back into the car and thought nothing more of it. It was at dinner with the group of farmers that evening that I realised when closing my right eye that I couldn't see a bowl of flowers on the table. Believing the problem to be only temporary, and recalling that I had been pulling ivy off the side of my house on Sunday, I assumed a speck of dust may be the cause of the problem. I drove home with difficulty and went to see the doctor, who sent me to Southend Hospital. After waiting for an hour I remembered I should be hosting a Riding for the Disabled event in Sandon.

I had been approached some time previously and asked if I would chair the local Riding for the Disabled group. Horse riding is an excellent therapy for children with health problems, and the position primarily involved attending a meeting every three months to discuss how sufficient funds could be raised to hire horses for disabled children to ride.

We needed thousands of pounds a year to hire the horses, and among the fundraising events we held was a summer fair, hosted in an empty barn belonging to one of the local farmers. There was also a special event to present prizes to the most improved riders. Held in an indoor riding school, I can recall one particular occasion when the moving nature of this event proved too much for one little girl, and she burst into tears.

I did actually make it through both my appointment at Southend Hospital and the Riding for the Disabled event, but at the hospital they referred me on to the specialist eye hospital,

Moorfields. At my appointment here it was discovered I had a retinal thrombosis in my left eye. The staff there were brilliant, and after a month of visits I met the surgeon in charge, who put a hand on each of my shoulders and told me to keep my head very still.

He asked me what sight I had lost, and I replied that I believed it was 50 per cent. He said he thought I was overestimating, and asked me to move my eyes left to right. I settled on 25 per cent, and just as he said, over time following the operation I underwent, in which they lasered the eye 2,500 times to prevent glaucoma, my one good eye slowly improved to the point where today I barely notice the lack of vision from my left. Focusing my vision on one eye also turned out to improve my shooting a great deal!

Other connections through farming and shooting proved very rewarding. In the mid-1970s a good friend of mine, Eric Spurrier, who worked in the gas industry with my father, rang me to say that his wife, a direct descendant of the Countess of Warwick, had inherited a farm. It was not far from Basildon, at Little Easton, near Dunmow, and comprised some 2,000 acres including the wartime American aerodrome. Most of the runways were still intact, but the farm had been very run down over the years. As with many airfields, most of the topsoil had been removed, and the heavy clay remaining contained lots of blackgrass and twitch. I did a lot of subsoiling on the airfield, and it was not unusual to hit buried concrete.

Years previously, when the estate belonged to the Countess herself, there were 100 acres of woods and beautiful gardens, which were now totally neglected and overgrown and ideal for a pheasant shoot. Eric asked if I would be prepared to run the shoot for him, put down pheasants, look after the feeding and provide a team of beaters on shoot days. It was not a big shoot – we were pleased to get fifty birds a day. Eric would invite eight of his friends and I would control the beaters. Among the guns was a lovely man, Ken

Corfield, who had at one time worked with Eric and my father but was now the boss of a very large telecommunications company. He used to provide and cook the most delicious steaks for our lunch on shooting days.

I was lucky enough to receive a number of invitations to shoot, and these were invariably of two kinds. Firstly there were those from the big farming companies who wanted to say thank you to their suppliers. These were great days, and one soon got to know who would be in the party of guns. None of us were brilliant shots as we didn't do enough shooting to become proficient, but we had a lot of fun and thoroughly enjoyed our days out.

Then there were the invitations to shoot by some customers who were hardly known to me. It took some time to understand the reasoning behind these, but they all tended to end the same way. Upon departure, I was often taken aside and told that one of the farm's tractors had recently required significant and costly repairs. After querying the age of the tractor and invariably establishing that it was over four years old and just out of warranty, I always said I would do my very best to see that it was fixed, arranging for the repair to be done under extended warranty. That sort of help usually resulted in an invitation to return to shoot the following year.

Ken Corfield often used to ring me in October and invite me for three days' shooting in Scotland, travelling from Luton in two planes – one for the guns and luggage and the other for the guests. We would fly to Inverness and then drive to Tulchan Lodge, a grand country house on the edge of the River Spey. After dinner we would use the bar in the sitting room and then play snooker in a room full of wall-mounted shotguns that had previously been owned by members of the Royal family, maharajas and other dignitaries.

There was a multi-national party of guests, including a Portuguese, three Belgians, an American and three Englishmen.

Breakfast was a feast: porridge, of course, followed by kippers, poached, boiled or scrambled eggs and bacon. Everything was first class, from the food and drink to the shooting and the company, and the trip included one of the bags of my life. The second day would be particularly special, as we would drive to the forest in the hope of seeing capercaillie, the legendary Scottish gamebird with a 5ft wingspan.

The beaters fired blanks to move them out of the trees, and on the first drive I was at the bottom of a hill, hoping the birds would not fly over me. The very first cock bird that broke cover, though, was heading straight over me. I fired both barrels but the bird appeared to fly on. The keeper came up to me, tapped me on the shoulder and suggested I'd never get a better chance of bagging my first capercaillie. I was mortified, but to my amazement one of the pickers-up appeared with the 12lb bird in his hand. It had dropped dead just 200 yards behind me. I must admit I was delighted, as none of the seven other guns even had a shot.

As a one-off shooting party this had been a memorable time, but it became an annual event for several years. So it wasn't all hard work selling tractors and dealing with problems – outside of day-to-day work we did have some very special days and this was certainly one of them.

I can't tell this Scottish story without mentioning my friends at Robertson of Tain in Ross-shire. Run by brothers Roddy and Kenny, theirs wasn't a big dealership by sales volume, but it was our most northerly, covering a vast area in the north-east of Scotland, including the Black Isle, and was a close neighbour of the Glenmorangie whisky distillery. Roddy was a great deer stalker, and three times I was invited to stay with him and go stalking. It was always beautiful in the mountains, but at its best when covered in snow. We wore white overalls as camouflage, and after going as far as we could in Roddy's Range Rover, used an eight-wheel-drive amphibious vehicle to cross rivers and get closer to the deer. It was vitally important to keep downwind to avoid the deer catching our scent. We used high-powered hunting rifles and did a lot of crawling on our bellies. When a deer was shot it was instantly gutted to avoid tainting the meat, and was then hauled back to the transport vehicle.

I recall one day heading up into the mountains in Roddy's Range Rover. It was pouring with rain, and when we reached the hydro-electric station at the top of the mountain, water was shooting horizontally for almost 30ft from 6ft diameter pipes before dropping into the loch below. It was an incredible sight.

It wasn't looking like a deer-stalking day, though, so we decided it wasn't sensible to hang around with the roads awash with knee-deep water and the river about to flood. However, as we progressed we realised it was almost too late, as rain had caused the river to flood the road to a foot and a half deep, just below the headlights. It was almost impossible to see the road, so Roddy told me to sit on the bonnet and point out the centre, doing my best to keep us on track. Eventually we arrived back in Tain, but here we met the young son of the shepherd who lived at the top of the hill from which we had just come. The nearest school was so far away that he boarded during the week and went home for the weekend, but on this occasion his father couldn't get down to fetch him.

Roddy offered to take him home, but having once escaped the floods, I thought we were mad to go back! I took my place on the bonnet once more, to guide us up the road. Luckily we made it through, and it was all worth it, with one little boy thrilled to get home to see his parents. We got back to Tain at 7 o'clock, five hours after leaving the mountains. Roddy deserved a medal for his driving ability.

7

Prairie Power

In 1975, John Deere brought one of its US-built 215hp articulated 4wd tractors to the UK, unveiling it to the public on its stand at that year's Royal Smithfield Show. While we understood it was done to gauge potential customer reaction, by the end of the event six had been sold. This was worrying, because each one had taken out at least four Ford tractors. However, in North America Ford had also decided to enter this market, through a deal with North Dakota-based Steiger, a specialist in articulated tractors. Four models from the Steiger range, spanning 265–335hp, were given a new blue and white livery and Ford FW-20/30/40/60 badging.

I went out to our training school in Texas, to assess these machines and see if they were suitable for pitching against the John Deere articulated tractors and others coming into the UK. In huge fields of 100 acres or more, I spent a week driving them, and found they had potential. They were what we wanted, although initially a key downside was that a hydraulic linkage was only optional, and although in the UK some tractors were likely to be used with implements such as trailed discs, we also saw a market for them working with draft implements such as large ploughs and subsoilers.

From Texas I went on to Fargo, North Dakota, where the Steigers were built. The temperature difference between the two states was remarkable, and while it could be very hot in Texas, it could drop to minus 40°F during January in Fargo. The factory was like a mammoth blacksmith's shop, with a lot of components made in-house. To give our customers a good deal, I ordered seven

265hp FW-30 tractors, which then had their fuel pumps recali-brated in the UK to give them the 295hp output of an FW-40, significantly more than John Deere's 8430 and 8630. The delivery period was six months, with transport alone taking two weeks on a train from Fargo to an east coast port.

On arrival in the UK, we had endless work to do on the tractors to get them 'homologated' to meet UK road and safety regulations, such as the installation of suitable indicator lights and a handbrake. This work was undertaken by specialist firm South Essex Motors, run by Eric May from premises not far from Basildon.

Launched at the Royal Smithfield Show in 1978, and priced at just over £40,000 on dual wheels, we nevertheless sold six machines for the following year. Van Geest Farms, of Spalding, Lincs, were our first customer, buying not one but two FW-30s for their large, flat fields. Our neighbouring farmer at Basildon, Peter Philpot, also bought two, while two others went to farms in Essex and Lincolnshire. We also had our own tractor for demonstrations and to act as a loan tractor in an emergency. Because of their width, with or without dual wheels, special low-loaders were needed every time we moved an FW on the road.

However, we soon found more than a few issues with adapting these tractors to UK working conditions, and I quickly had a bit of a nightmare on my hands. The first phone call I received was from Peter Philpot, to say that one of his tractors had a fuel leak, with diesel not just dripping but pouring from a cracked fuel tank. The linkage was designed with the top link connected to the bottom of the fuel tank, and this wasn't robust enough to cope with heavy draft work. Eric May welded the cracks, fitted a new bottom plate and relocated the top link bracket, after which all seven tractors then in the country had the same treatment.

Everyone was very quick to tell me that, as it was my idea to bring the FW tractors to the UK, it was up to me to solve the problems! With a great deal of help from the unflappable Eric

May we got through the first year and had a number of modifications ready for year two. I was greatly relieved that we imported only those seven tractors during that first year. I, the company and the farmers who purchased the first six FW-30s – and later models – greatly appreciated Eric's engineering help.

We also worked hand-in-hand with Roger Dowdeswell, another brilliant engineer, whose Warwickshire firm made the huge reversible ploughs of up to ten furrows to suit the FWs. In dry weather they worked fine, but as fields got wetter the tail wheel on the plough, which had to be reversed when turning on the headland, used to sink, causing the lift arms on the tractor to scissor and be pushed upwards beyond their limit and break in half. If the linkage was set to position control, the tractor had no draught control. Somehow, we survived that year, and for 1980 ordered 16 units.

Demand continued to grow as farmers with big fields saw the potential to cover large acreages rapidly with just one man, and cultivate deeply to allow better rainwater infiltration and crop root growth. While some cultivator makers designed wide sets of discs and other implements that folded for road transport to exploit these tractors' power, others sought to achieve the same by creating implements that were not necessarily wider but longer, carrying out multiple soil loosening, cultivation and consolidation operations in one pass.

With the demand for greater workrates seemingly unfulfilled, at the 1980 Royal Smithfield Show we launched a second model in the UK range, the 335hp FW-60. One of the first sales I can recall went to a large farmer in north-east England, whose neighbour had recently purchased an FW-30. The farmer decided he also needed an FW, but being keen to play a bit of one-upmanship with his neighbour, asked if he could possibly have a larger machine. As we had recently launched the slightly bigger tractor, he was delighted to place an order guaranteeing him a larger machine than

A photo taken during the earliest days of my career with Ford, aged 21 in 1957 and in Cuba on tour with the demonstration team launching the Fordson Dexta.

In the driving seat of the Fordson Dexta, learning the art of disc ploughing during our demonstration tour of Cuba.

Enjoying some Cuban hospitality with two of the locals while on our demonstration tour of the country.

A picture of me aboard the Dexta at its launch at London's Alexandra Palace in 1957 was used across the farming press, and in the Farmers Weekly alone was printed 25 times!

Taken some time in the late 1950s, this photo shows me cutting the lawns at Boreham House using a Dexta and a set of gang mowers, probably made by Ransomes.

My team and I ready to set off from Boreham House on our European demonstration tour of Belgium, Holland, Sweden and Finland.

With the demonstration team, planning our route around northern Europe for our Dexta demo tour during 1959.

Taken in 1959, here I am leading the way on tour with the Dexta demonstration team in Omagh, Northern Ireland, shortly after the tractor's launch.

Another photo taken when we were en route between demonstration sites during our Northern Ireland tour, with me in the lead tractor towing a Jones baler.

Launching the new Power Major in front of our dealers and the press at an event held in the grounds of Boreham House.

Seated aboard the new Power Major in a marquee at the dealer and press launch held at Boreham House in 1958.

In 1960, the Super Major brought with it a number of welcome developments, including an all-new draft control system, plus a differential lock and disc brakes.

The opening of the new Basildon factory in Cranes Farm Road coincided with the launch of the 6X series, the new 2000, 3000, 4000 and 5000 models that ultimately became known as the 'Pre-Force' range.

Although the Ford logo is long gone and the sign has changed to read New Holland, today the frontage of the Basildon plant remains largely unchanged. This photo was taken not long after the plant's 1964 opening.

A view to the right of the factory grounds at Basildon, with the 'onion' water tower in the background. Among the Pre-Force/6X tractors parked is an export version with outboard headlamps (bottom left).

Launched in 1964, the new 6X series tractors retained their Dexta, Super Dexta, Major and Super Major names alongside the new 2000, 3000, 4000 and 5000 numbering.

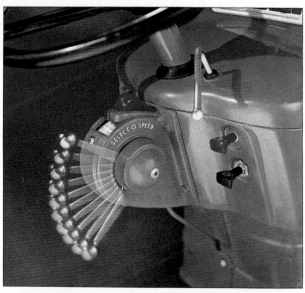

Introduced on the 'pre-Force' 6X tractors, and also featuring on the later Ford Force 6Y models, I rated Select-O-Speed as a wonderful gearbox, but reliability was a major issue, primarily because of a lack of understanding in how to operate it.

In 1968, Ford chose Southend Airport as the venue for the dramatic launch of the new 6Y 'Ford Force' range, flying the tractors in aboard a transport plane before driving them off in front of the assembled dealers and press.

At first, the 6Y 'Ford Force' tractors were supplied as less-cab models or with after-market cabs from the likes of Duncan and Lambourn. Ford enjoyed a strong marketing partnership with Ransomes for many years, as illustrated here.

A Ford 5000 making silage with a Wilder Sila-Masta forage harvester loading into an FW Wheatley trailer. The 75hp 5000 became one of the most popular of the 6Y tractors.

It was during the early 1970s at Boreham House that we trialled an after-market turbocharger from Northrop on a Ford 5000, creating the first 94hp 7000. This tractor also introduced the Load Monitor hydraulic system to the market.

In late 1975 Ford introduced the 7A1 range, or '600 series', to succeed the 6Y tractors. The new Q cab design was delayed and the new machines were launched with the old Fieco safety cabs.

The 1976 quiet cab regulations saw the introduction of the new Q cab for the 7A1 tractor range. The fact it was designed with two doors, with easy entry/exit possible from both sides, was a strong selling point over rivals, particularly Massey Ferguson.

While it had long been in the industrial equipment market with highway tractors and loader-backhoes, Ford's 1979 purchase of French construction machinery firm Richier saw me become involved in the marketing of wheeled loaders and tracked 360-degree excavators.

Taken in 1977, this picture shows me making a presentation to Ted Gates (right) of Hertfordshire tractor dealer Gates of Baldock, to mark the company's 50 years as a Ford agent.

In conversation with agriculture minister Michael Jopling (centre)
and Geoff Tiplady (right) at a show some time during the early
1980s.

I went out to test drive the Ford FW tractors in Texas and visited the Steiger factory where
they were made before deciding to offer them for sale in the UK. The Cummins V8-powered
295hp FW-30 was launched at the 1978 Royal Smithfield Show, joined in 1980 by the
335hp FW-60.

It's a grainy photo, but this picture of the Basildon factory from the early 1980s is a very evocative one, with two Series 10 tractors aboard a Ford Transcontinental truck, and the 'Onion' water tower in the background.

The Series 10 tractors were launched in 1981, with one of the main – and ill-fated – upgrades being the little-liked column gear-shift for the 7610 and below, designed to create a semi-flat floor.

Ford was very keen on the publicity earned by attempting and breaking records, such as this early 1980s 144-hour ploughing feat using front and rear reversible Ransomes ploughs mounted on a Series 10 7710.

In deep conversation with the Duke of Gloucester at the Royal Smithfield Show during the late 1970s, discussing his tractor requirements.

Members of the Royal Family were frequent visitors to the major agricultural shows. Here I'm pictured with Princess Anne at the Royal Smithfield Show in the late 1970s.

Sharing a joke with Princess Michael of Kent as she tries out a Ford 1210, our smallest tractor, at the Royal Smithfield Show some time in the early 1980s.

Another photo that's more than a little grainy, but it gives a great idea of how our show stands looked in the early 1980s. Among the tractors pictured on this Royal Show stand are a 1200 compact, a 6610 4wd and, of course, an FW.

Launched at the 1987 Royal Smithfield Show, the 98hp six-cylinder 7810 resulted from work begun by Eric May at South Essex Motors. Ford purchased the design from SEM, and in its first year the tractor made a profit of $6 million, becoming one of Ford's most popular sellers of all time.

In 1989, to mark 25 years since the opening of the Basildon factory, a Silver Jubilee version of the Generation III 7810 was unveiled at that year's Royal Show, complete with a comprehensive specification including many extras. While they created relatively little interest at the time, later the tractors became collectors' items.

Also shown at the same 1987 Royal Smithfield Show at which the 7810 was launched was the 276 version of the nearly identical Versatile 256 Bi-Directional tractor shown here. With a reversible operator console, it could be driven facing either direction and featured a linkage and pto at each end. After testing, we decided it wasn't a viable prospect in the UK.

The 1991 Royal Smithfield Show saw the launch of the new Ford Series 40 tractors to replace the Series 10 Generation III models. The 5640–8340 superseded the 5610–8210, and used the new Ford Genesis engine.

One of the last major projects with which I was involved was the development of the Series 70 replacements for the Series 30 tractors that had succeeded the TW models. These were a clean sheet design built in the Winnipeg, Canada, factory we had acquired with Ford's 1987 purchase of Versatile.

I helped to launch the Series 70 tractors in a number of markets, and particularly recall events in New Zealand, where I met up with my very good friend John Parfitt, standing next to me here, who managed the country for the company.

his neighbour. In 1982 we sold around 48 units of the two tractors, which was more FWs than Ford sold in the US that year. But by that point we had saturated the UK demand for big power, and over the following years sales dropped to a handful annually.

Powered by V-8 Cummins engines, the FWs were noisy tractors, and this caused particular problems at night when people were trying to sleep, as in order to work at maximum efficiency nearly all owners worked their machines 24 hours a day. One summer night a school master, trying to sleep, telephoned Peter Philpot at his home at 2 o'clock in the morning and suggested that if he didn't stop his tractor working he was going to go out into the field and lie down in front of it to make it stop. Peter suggested this wouldn't be a good idea as the driver's short-sightedness might result in him being run over! In the end we found a very sophisticated silencer, which was mounted on the fender and reduced the noise level dramatically.

There was a very definite pattern to FW sales, following the heavy land farms across Yorkshire, Lincolnshire, Cambridgeshire, Suffolk, Essex and Oxfordshire. Many owners told me the high-output tractors and associated equipment had revolutionised their farming. A total of 145 FW-30 and FW-60 tractors were sold between 1978 and 1988. I feel sure they will one day be collector's pieces, as is the Doe Triple-D, and indeed some are now being restored and shown at rallies and tractor gatherings.

One of our biggest customers for FW tractors was CWS, the farming arm of the Co-op supermarket group. Sometime in the mid-1980s I took a phone call from Alan Dodsworth, national farm manager for CWS, which at that time was one of the biggest farmers in the UK. It ran about 200 tractors on its farms across the country, covering all different makes including Ford, but wanted to standardise on one manufacturer.

He asked if I would prepare a presentation to the Co-op management and directors, to be held at the head office of its farming operation, which was in Leicester. We met in due course and went through the whole range of Fords, from narrow small tractors to work on its fruit farms to the 300hp+ FW articulated machines. The availability of Ford dealers adjacent to its large farms up and down the country from Essex and Gloucestershire in the south to Yorkshire in the north was a key factor, I suggested, and we then discussed our fleet owner discount scheme, which certainly made them sit and make notes.

No decision was to be made that day, and a follow-up meeting was arranged. Within one month I was back in the office for another session, and this time we talked about second-hand values of tractors, for which I had prepared a list of documented used values of Ford and Massey Ferguson models. With the presentation finished, they asked me to leave the room and said they would call me back in a few minutes. This they duly did, after which they announced that their choice was Ford. I was delighted – it represented the biggest ever tractor deal in Ford's history.

Mike Calvert, a very capable and nice man to deal with, was the boss of Co-op's farming business, but primarily I used to work with Alan Dodsworth, who ordered new tractors at regular intervals, and he became a really good friend as well as a customer. I used to get a phone call from him saying he would like to order ten 8210s and an FW-60. I would then programme the factory build and would be told which Ford dealer had got the deal.

With farms up and down the country, the Co-op purchased not only a huge variety of tractors, but a huge number, too. I can remember it buying eight FWs and probably 100 8210s, and even Ford orchard conversions for its fruit farms. It was also very generous, and used to invite its major suppliers to shoot with it once a year at its Cirencester farm, where we always had a wonderful day, preceded by an excellent night before the shoot.

I recall the day's last drive being particularly spectacular. The guns lined a clearing in the wood and the beaters walked past us, driving the birds ahead of them until the end of the wood, when they all flew back over our heads. It was a very special day and one we enjoyed immensely.

All worked well while I was general sales manager, but one day someone in Ford's finance department decided to change the fleet owner discount and break the promise I had given them in writing. The Co-op never purchased another Ford tractor.

8

Excavators and Exhibitions

Having spent two years as southern area sales manager, helping regain market leadership for Ford, in 1979 I was promoted to UK marketing manager. It wasn't a job I especially relished, as my real interest was selling. I hated having to talk about advertising, preparing for the Royal Show and deciding who got the contract to build the Smithfield Show stand. But I got stuck in and I survived a couple of years before moving on to a new position, more of which later.

I had hardly been in the job a week when it was announced that Ford had decided to buy the bankrupt French construction equipment manufacturer, Richier. A few days later a group of us flew to Paris, chartered a private jet and arrived at the Richier plant. The firm seemed to make everything – we looked at tracked 360-degree excavators, a range of wheel loaders, mobile concrete mixers, tower cranes, road rollers and compactors. In the end we decided to focus on the tracked and wheeled 360-degree excavators, a range of three wheel loaders, and a line of tower cranes.

The latter were really scary contraptions. The first problem with them was with bolts shearing in the boom, with the danger that the whole crane could collapse in a heap – although fortunately this never happened!

The best products were the wheel loaders with their automatic transmissions. They were very easy to drive and had a good reach, making it possible to drop the load in the middle of a truck. I loved driving these loaders and it wasn't long before I was making my first presentation of this product to our industrial dealers. We had 16 specialist dealers for industrial equipment, some of which were also tractor dealers but others who focused solely on the construction and associated markets, with depots located in cities such as Glasgow, Cardiff and Leeds.

The 360-degree excavators were also very good, the biggest being a one cubic yard machine. I remember Jones Brothers of Ruthin in north Wales selling one that was to be used to make the power station in the Welsh mountains where the water ran through huge tunnels cut into the rock, to fall down into the turbines before being pumped up to the top of the mountain to repeat its journey. These tunnels were huge – big enough to take a double-decker bus.

I enjoyed taking any opportunity to operate the excavators, and one day I offered to help a friend clean out his garden pond. I was doing quite well when the digger started to sink into the pond. I tried to push my way back up the bank but the bucket was sinking into the mud at the bottom of the pond, and I was having visions of the whole excavator disappearing. A call to my good friend Malcolm Denton saw him arrive quickly and help me out, while suggesting I should give up my earthmoving career before I got into real trouble!

The Richier acquisition meant that Ford had become a serious construction and earthmoving equipment firm as well as an agricultural tractor one, and we needed a dealer organisation to match, beyond our existing agricultural dealer network that sold a few tractors for construction use. This proved a challenge, and we often had to approach people who were involved in the industrial business hiring out machinery to the end-user. It was one of those companies that I went to see in Glasgow.

We had agreed to meet the management in their yard at 9.30am, and our industrial sales manager Keith Banks and I agreed to leave at 4am. However, a stroke of luck meant I was able to obtain far faster transport than we had originally planned to take. The night prior to departure I had a meeting with Sir Leonard Crosland, who was a director not only of Ford but also of sports car specialist Lotus. I told him about my proposed trip and he suggested I take his Lotus Elite. This was a very special car, and Keith and I soon found it was capable of cruising at 140mph, far faster than I had ever been before in a car. We arrived at the proposed new dealership before they had opened the gates, and had to wait an hour for the management to arrive.

My only other noteworthy experience I can recall with the Richier equipment we had acquired was with a man who had a wheeled excavator equipped with pneumatic hammer and breaker. He was very pleased with his machine, and suggested to me he could break any concrete structure. I introduced him to my farmer friend, Tony Speakman, who, as a challenge, had a wartime pillbox in the middle of a field.

The customer was convinced the job would be no trouble, but after a day of continuous hammering and broken points the pillbox was still in the middle of the field – where it remains to this day!

Ford soon found that the industrial market was highly competitive, and we had to give huge discounts to make a sale, which meant very little profit at the year-end. Buying Richier may have seemed like a good idea at the time, but ultimately we were well out of the heavy end of this market. Ford later scaled back its involvement in construction equipment to focus on tractor loader backhoes.

After a brief period as UK marketing manager, I achieved a goal I had never dreamed of, but in some very sad circumstances. At the time, Laurie Chivers was our director of sales, while Geoff Tiplady was general sales manager, and one particular day in 1977, Geoff and I had a phone call to go and see Laurie at his home. He told us he had cancer and would be retiring that day. Laurie asked Geoff to take his job and me to take Geoff's. It was a sad day for us all, but work had to go on.

As general sales manager I was responsible for all sales, service and the supply of parts. I had two very good area managers, one of whom was Arthur Hebblethwaite, who looked after the north of the country. He had been very good to me when I joined the company, and was an amazing man, very good at managing people. One day he was summoned to see a castle-dwelling Scottish laird who had decided due to some issues with his Ford tractors to replace the entire fleet with another make. Arthur entered the office to find a scowling brigadier sitting at his desk, and noticed a photograph of his old army regiment on the wall behind him. He pointed to the photo and said:

'The colonel there – he's my father-in-law.'

The brigadier was amazed, got out the whisky and the subject of tractors was never mentioned!

Harry Watson was the southern area manager, an Irishman from Cork. He was very popular with all his dealers, as I observed first-hand as we often travelled together to visit them. Harry was a wonderful friend and always did his best to keep me away from problems. It was a very sad day for me when he retired.

There were eight sales zone managers, who were located on their respective territories, while on the service side Keith Briggs was our service manager. His career was cut tragically short when he suffered a completely unexpected heart attack and died in a Norfolk hotel. His job was taken by Peter Edwards, who was in charge of seven service zone managers, also territory-based, looking after the tractor range and the tractor-loader-backhoe (TLB) product.

I was always prepared to talk to farmers on the phone and had two phones in the office so no one was kept waiting, always agreeing to call them back if I was engaged on the other line. It was hopeless ringing back at, say, 11am, as the farmer would be back in the field, so I decided always to call back after 6pm. They were always surprised to get my call and more surprised that I was still in the office. I used to start at 7am so I could get work done before the phone started to ring, and never left before 7pm.

However, it was much more interesting to meet customers in person, and if this wasn't on their own farms, then it was invariably at one of the national or county agricultural shows. Held each December until its later years, when it went biennial, the almost week-long Royal Smithfield Show was the highlight of the farming year for many. Farmers from all over the country used to descend on Earl's Court in London, and if a manufacturer had a new product to unveil, they would almost invariably do it at Smithfield.

There was little chance of going round the show once the farmers had arrived so I used to have an early breakfast and arrive at Earl's Court at 8am. I started by touring the stands of our key competitors, before moving on to the lesser-known makes the following day. It was all very friendly, and as I knew most of the staff on the tractor stands they were always keen to show off any improvements they had made to their machines, big or small.

It was also an excellent place for customers to provide us with feedback and discuss any issues they had. One morning I had just walked on to our stand at Smithfield, and hadn't even had time to put on my badge, when a farmer, recognising my blue blazer, asked me for some help.

After I had introduced myself, he exclaimed:

'I don't believe it. I've just travelled down on the train from Lincoln and there was a group of farmers in my carriage. We got talking and they said that if you want any help from Ford you must see David Pearson. Now here I am talking to you within two minutes of arriving at the show!'

It was one of the nicest things ever said to me by a customer, which is why I recall it to this day. I sorted out his problem there and then, making note of his name and address so he could be contacted as soon as the show was over to help with the issues he had. He went off a happy man.

Smithfield was very important to me, and I very seldom got a break, certainly never taking time off for lunch on the first three days. One bit of particular excitement I recall from the mid-1970s was a visit to our stand by HRH Princess Michael of Kent.

First of all I showed her one of our tractors with the then-new flat deck cab.

She asked if she could take a seat in it, and I opened the left-hand door for her. She climbed aboard and sat down, but then, when deciding to depart, opted to do so from the right-hand side, where the gear levers were. She was wearing a full length skirt, which engulfed them both, and promptly halted, informing me of the problem! I politely suggested she continue her progress, and thankfully she did so with all ending well! Each time I come across the photographs I smile.

Just as important were dealer open days and events, the largest of which was the annual Doe Show held by our dealer for Essex and the surrounding areas. Every year, in the first week in February, after the shooting season had ended, Ernest Doe and Sons would use the land surrounding its headquarters branch at Ulting, near Maldon, to show off the latest equipment from its franchises, as well as sell off used machinery. This covered a huge range of equipment, from the biggest combine to the smallest lawnmower, and there would also be working demonstrations of tractors and a full range of industrial equipment from tracked diggers to chainsaws. The show lasted for three days, and was held in every type of weather, from bright sunshine to snow flurries and torrential rain. February would normally be a quiet month for the dealership, and for farmers, but the show would produce large numbers of sales.

Alan Doe, the company's managing director and latterly its chairman, would produce roast beef lunch for a dozen customers and staff from key franchises at his house next door, all cooked by his charming wife, Jill, and on many occasions I was invited to lunch with them. Hundreds of people attended the show, coming down in coaches from beyond the Doe trading area to seek out a bargain and get a closer look at every blue tractor, from the smallest compact to the biggest articulated machine. From the warmth of Earl's Court in December to the chill of Essex outdoors in February, these shows and events and the chance to meet customers were some of the most enjoyable aspects of the job.

9

Dealer Challenges

In East Anglia, for many years the main Ford agent across Norfolk and Suffolk was Mann Egerton, which had three tractor dealerships at Norwich, Ipswich and Bury St Edmunds. Run by John Hart, it was one of the best Ford dealerships in the country.

However, we were informed in the mid-1980s that the company had decided to close the Bury St Edmunds depot, and this left me with the task of recruiting a new dealer.

There wasn't a lot of choice in the area, but there was a very good long-established International Harvester dealer called Cornish and Lloyds, in the same town. I did some research and found it had been in business a very long time, with a long history as an implement manufacturer, previously manufacturing a horse plough that was not unlike that made by Bentall. During the depression in the 1930s, the firm had a clever sales plan by which it would give any farmer who purchased a dozen plough shares over time a single-furrow Cornish and Lloyds plough in return. The company also made Cambridge ring rolls in its foundry, well into the 1980s, which closed around that time as the economics of buying in versus forging its own rings made it unviable.

I arranged a meeting with Mr Chapel, the managing director, but we didn't get off to the greatest start when he told me he didn't really trust Ford Motor Company, suggesting we had a reputation of delivering tractors without an order and then demanding the money for them. He was a careful man and if he found a spare bolt on the floor he would take it back to the stores and have it re-binned. Likewise, if something was delivered to the dealership and not wanted, it was immediately sent back and a credit note requested.

I wasn't doing well in persuading Mr Chapel to change his franchise from International Harvester to Ford. However, I did discover that Bill Mann, who owned and ran Manns of Saxham, the Claas importer for the UK, was a shareholder in Cornish and Lloyds. I arranged a meeting with Bill, who I had met a couple of times when a shooting guest of Ernest Doe. We talked a lot about the benefits of the volume sales potential of Ford tractors when compared with International Harvester models, which were less popular. He was very receptive and said he would see what he could do for me to help get Cornish and Lloyds to change its franchise to Ford. We had many meetings and in the end it came down to the added spare parts business. When I had finished my final meeting with Bill, he walked downstairs with me to my car and I departed. It turned out however, that we were seen shaking hands because at exactly the same time a car load of sales managers from Ford and Claas dealer Tildesley were driving in. Why was David Pearson talking to Bill Mann? Anyone with a vivid imagination could quickly come to the conclusion that Ford Motor Company was going to buy Manns of Saxham. The rumour spread like wildfire and it was fully a week before we could tell anyone there was no proof in the story! I am pleased to say, that with Bill Mann's help, Cornish and Lloyds became good Ford dealers, eventually being bought by the Dalgety organisation, under whom they continued to sell our products.

Not all our dealer moves were good ones, though. The shake-up following the Case IH merger had left some good dealers without franchises, such as Platts, in the East Midlands. However, the decision made higher up in Ford during the early 1990s to drop Sharmans, our successful dealers for much of Lincolnshire, and give the franchise for the area to Platts, which was owned by the car dealer group Lookers, seemed the wrong one to me. While

Platts became good dealers for us, Sharmans went on to sell even more tractors, having quickly been picked up by John Deere, who soon became market leaders.

***

In 1983, the original TW range was replaced by new TW-15/25/35 models, with uprated engines, greater lift capacities, and a new design for the middle model that gave it the extra fuel capacity and longer-nosed look of its big brother. In 1985, along with the Series 10 tractors they were then given the new Super-Q cab, with its square looks and bank of worklights. But with only a conventional synchromesh transmission with Dual Power splitter, Ford was lagging behind on the big tractor transmission front.

One day I took a phone call from Harris of Epworth, our dealer in north Lincolnshire, saying that one of its biggest customers was about to change his whole fleet of Ford tractors to another make, and asking if I would come and see him. I dropped everything, got into my Lotus Cortina and headed up the A1. There were no speed cameras in those days, so I could push on, and I arrived at Epworth in plenty of time. I was told by the dealer we were going to meet the farmer at 11am the next day, which gave me ample time to go through all of his problems with the Harris staff. The latest and most important was a TW-25 well out of warranty with a broken crown wheel and pinion.

The next day, we arrived at the farm office dead on time and met the boss and his two sons. The farmer was very angry, to say the least, but we gradually worked our way through all his problems. This, though, took most of the morning, and as half past one came round, I thought that, even given the hostility between us, we might be going to lunch. As if reading my mind, though, at that moment he tipped his wooden chair back, jamming the door to the office. I had to do something if we were going to get out of the office before he picked up the phone and invited a competitor in.

The Ford 8210 was fairly new on the market at that time, and I asked if he had tried one, which it turned out he hadn't. I told him I could arrange for a new one to be delivered to the farm on a six-month free trial, and if he liked it he could buy it for half price. As for the TW-25, I would give him an ex-demonstration unit, with very low hours, in exchange for it.

This offer seemed to placate him, and after a bit more haggling he agreed to do the deal and stay with Ford. He left the office and we sat down with his two sons, who then told us some stories about their father. It turned out he had a sentry box at the entrance to the farm and if one of his men arrived two minutes late in the morning he sent them home for two days without pay, suggesting they must be too tired to work. The other brother then told us he had a bottle of champagne permanently in his fridge, which he would open on the day his father died.

I wrote the report on my visit the next day and sent it to Geoff Tiplady and to Mike Dyos, who worked on the finance side of the business. Geoff called me to his office and asked if I had gone mad, but Mike came to my rescue, persuading Geoff it was a good deal bearing in mind the number of tractors involved. My attitude to problems was: 'soonest mended, soonest ended'.

As it happened, the gearbox on the TW was the most reliable part of the tractor, but it had seen better days, and during the late 1980s our engineers in the US were working with the German company ZP on developing a new powershift transmission – although initially we in sales knew little of this. Sometime after development had begun I was invited to visit the test farm in Georgia in the US to try it out. By now I had learned that I needed farmer back-up on one of these trips, to give practical feedback – the engineers did not like me telling them their baby was ugly, but they would listen to end users!

I had recently met up with two of the most switched-on farming men I came across in my career. Richard Dodds was manager of a very large farm in Hungerford, Berkshire, while Peter Slugget, a contractor, was from Devon. I invited them to come with me to test the new transmission, and also took a good friend from Ford, Malcolm Denton, who was a zone manager based in Oxford.

I insisted on business class tickets for our VIP customers. I don't like to remind Peter of this, but I recall sitting with him on the plane and him saying that the first thing he must do when he arrived at the hotel was to ring his wife. I reminded him of the six-hour time difference, and that it would be 3am in England upon our arrival. He obviously didn't believe me, because when we came down to breakfast he said she was furious at being woken at that ungodly hour!

We had three prototype 18-speed powershift tractors available to us in the field, and the transmission had two ranges, with clutchless changes from gears one to nine, and then a manual lever change before the second range of gears from ten to 18 was accessible. The manual shift occurred at a forward speed of about 5.6mph, meaning that to get to 6mph and beyond it was necessary to stop the tractor and change ranges, losing inertia and making smooth progress difficult. After a day in the field my guests and I had decided that for the European market it was a non-starter. We elected Richard Dodds to be our spokesman, and he did a brilliant job of telling the engineers that their baby was really ugly and he would never buy one!

Things were a little awkward after that. A special dinner party had been planned for the evening to celebrate us saying yes to the new programme, but before it began the chief engineer had gone to the airport and flown home, not a happy man.

The design was dropped, and Ford began working with Funk, an American firm that specialised in transmissions for large earthmoving equipment. About six months later the whole team from the UK returned to San Antonio, Texas, to try out the new powershift. It did have a glitch between two gears when all the hydraulic clutches were working, but Funk was able to rearrange the gear shifts to avoid this problem, and compared to the original powershift it was brilliant. The TW-15/25/35 tractors were relaunched in the UK at the 1989 Royal Smithfield Show as the 8630, 8730 and 8830, complete with this electronically-controlled 18 × 9 powershift with a short, stubby lever that was moved forwards or backwards to select the direction of travel and tapped right/left for up/downshifts.

The Funk people gave us a 30kph (20mph) top speed for Europe, while North American models made do with a 25kph (15mph) maximum due to legal requirements. There were some clever engineers within the firm, and they also gave us features such as an automatic up and downshift based on tractor engine rpm.

10

Contracting Capers

To keep my hand in with practical farming, and to help pay for my children's education, during the 1980s I rarely took a holiday, instead doing a lot of harvest contract work during my summer break and at weekends. I did some mole-ploughing with an FW-30 on the farm owned by Sir Leonard Crosland, and worked for a time for Co-Partnership Farms, but mainly I would work for my very good farmer friend, Tony Speakman, who always needed extra help at harvest time preparing the ground for the next year's crop. To take these opportunities I used to take four weeks' holiday in a block, starting the day the first field of corn was cut.

I would find a TW-30, TW-35 or 87/8830 demonstrator, and could normally get the tractor for a small deposit, with the majority on hire purchase for a period of one month, after which I would usually sell it to a friendly dealer before I had even purchased it. To go with the tractor I would buy equipment as and when I saw a bargain. Over the years I purchased – and sold on quickly for a profit – a five-furrow Kverneland reversible plough, with a press to match, a trailed mole drainer, a 6m set of discs and a 6m cultivator.

My main task was ploughing, and I liked to start early in the morning, beginning work at around 5.30am, to pack in a long day. I can recall once slipping when exiting the tractor, and spending the rest of the day in pain until I had my wrist examined later that night, when it was found to be broken.

The Kverneland plough I purchased was a Vari-Width model capable of ploughing at furrow widths of 12–20in. I was very fond of ploughing and I liked to do it right, but when I ploughed right-handed the third furrow always ploughed high. One day I bent a leg on the plough and took it off to get a replacement, only then discovering that the three holes that had been drilled in it to attach it to the mouldboard were about an inch out of line. Kverneland was very good to me and gave me one of its demo units at a very reasonable price as an apology for the error.

Tony's eldest son, Richard, who had now taken over the running of the farm, always looked at the pointer on the plough that told him the furrow width. When I was ploughing at 18in he would ask me to go back to 12in, reducing my daily acreage. I knew what the tractor could do, though, and decided to reset the pointer so it read 12in but it was in fact working at 18in. Everyone was happy!

I extended my fleet of implements by purchasing a three-leg subsoiler through my good friends at Ransomes. To create a more challenging implement for a lower cost, I then had it welded on to the back of another subsoiler frame and fitted six extra legs, making it a total of nine legs. It made an excellent job, particularly when I fitted a crumbler roller on the back.

One year I also boosted my work capacity by buying a 154hp TW-25 and then purchasing a second-hand 186hp TW-35 engine from the field test department, relishing the challenge of swapping the engines. Unfortunately I neglected to change the crown wheel and pinion, and while this never proved to present me personally with a problem, two years later a rather irate farmer who had latterly become the owner of the 'hybrid' tractor suffered a breakdown in the rear transmission. After checking the log book he discovered his machine was really a TW-25 and someone had changed the decals! I found him a new crown wheel and pinion and managed to placate him though.

One of the other implement companies with which I used to work closely was Howard Rotavator. Following the success of the Rotavator itself, it designed another rotary cultivator, but this time equipped with paddles rather than blades, which the company lent to me to field test. I did hundreds of acres with it, and felt it was a first class topwork tool. Howard, however, decided not to go into production with it, and sold it to me at a knockdown price.

Howard also produced the Paraplow, which used disc coulters ahead of angled subsoiling tines set along a plough-type beam. On grass its purpose was to aerate pasture without undue disturbance of the surface. It was a very good implement and worked well, but ate through steel, and as a result was very expensive to run. To my surprise, one day I got a phone call from a competitive manufacturer saying it wanted to buy my unit with the idea of manufacturing the machine. I doubled the price I paid and said the company could collect it at its convenience.

During my contracting period, I recall once taking a call from a friend saying he needed a field ploughing as soon as possible, and asking if I could help. I said I could do it the next day, and he was delighted. I'd ploughed half the field when I noticed what looked like a golf ball on the ploughed land. I had no idea there was a course on the other side of the hedge, but at that time I was playing a lot of golf and was pleased to have the balls. Very soon all my pockets were full and I had a bag in the cab with close to 100 in it.

I returned to this gold mine a week later with my gundog, but she didn't appear to be interested in picking up for me. On the way home with my dog beside me, though, I noticed she had a golf ball in her mouth. Hoping I could train her up as a golf ball hunter, I returned a while later with the dog and a bag of Maltesers as a reward for when she picked one up. She very quickly got the hang of it, finishing off all the Maltesers and leaving me with pockets full of golf balls. Together we found about 100 each time we went searching, although at these sorts of numbers, I should point out, she wasn't given a treat for each find!

During the early days of the 8030 series I had my own 8830, purchased through Ford Credit and run for three months before selling it. As part of the transmission development I had four black boxes in the back of the cab, all controlling different speeds and different ratios together with the automatic up/downshift, which was handy for roadwork. It was then the only development tractor of its type in England and it used to scare the life out of anyone who tried it – as long as the engine was running at over 1,600rpm the transmission automatically changed sequentially up to 18th gear. This was one of the last tractors I ran before I drew my contracting sideline to a close. It had been a great help to my finances over the eight years or so that I decided to do it, and helped to pay for my children's education, but they and my wife preferred to see more of me over the summer!

11

Successes and Failures

The mid-1980s was a great time to be general sales manager, during a period when Ford's agricultural business was in significant expansion. In 1986, a year after the Force II update for the Series 10 and TW tractors, which included the new Super Q cab, the company purchased the New Holland combine and forage equipment business. Just a year later it went on to purchase Versatile, the Canadian high-hp articulated tractor specialist. Ultimately, the purpose of these acquisitions was to launch the new full-line farm equipment business, Ford New Holland, as an independent entity on the stock market. Farming, though, was in recession, and so we went it alone for the time being, while Ford, which had decided its future lay in focusing on cars, spent the next few years looking for a buyer for its farm and construction equipment interests.

Following the New Holland acquisition from Sperry Corporation, at Ford we set about the process of integrating the two organisations by first interviewing New Holland's UK-based staff, introducing them to our tractors while we familiarised ourselves with their combines, balers and forage harvesters. It was a whole new world for all of us, and not only did we have a cohesive product line to create, but there was also a dealership network to reorganise. In time we had a new dealer organisation, although initially not all took on both lines.

***

When Ford Motor Company purchased Versatile in 1987, it inherited not only its range of big four-wheel drive articulated tractors up to 470hp, but also the 276 Bi-Directional, or Bi-Di. This was a small articulated machine of 116hp with a cab at one end, a reversible driving console and linkage/pto at each end, designed to be driven in whichever direction suited the job. When operated conventionally the driver looked out over the bonnet, although more unusually the articulation point was in front of him. When working in reverse-drive mode, the operator looked immediately out to the rear, and many North American farmers used the Bi-Di to operate a rear-mounted loader in this manner, or a mounted swather or mower. There were, though, a couple of peculiarities that didn't suit it particularly well to a wider world market working in different conditions. The transmission was hydrostatic, which meant it wasn't great for draft work, while the front and rear linkages offered no draft control.

However, American management considered the Bi-Directional to be the plum in the pie when Versatile was purchased. Being a Versatile product it was bracketed with other high-hp tractors, and as product manager for these, I was given the job of evaluating its potential for sale in Europe.

Two units were delivered to the training centre at Boreham House, each complete with a rear-mounted loader, and the first thing we did was to hook one up to a small three-furrow reversible plough to assess its fieldwork credentials. Being hydrostatically-driven, it proved impossible to control the tractor's forward speed. It went well in very light land, but in heavy ground it came to almost a standstill, even with the engine at full rpm. We wrote it off as a ploughing tractor and fitted the loader. This was a simple operation involving four pins and a couple of hydraulic pipes, and the loader worked quite well, the driver being located just behind it. Again, though, the hydrostatic drive would make the tractor jump forwards as soon as the bucket rose above the material being loaded.

I should mention that the tractor was a good seller in North America, with many farmers using a Bi-Di in the long winters for clearing roads of snow and feeding cattle, and in the summer working with a swather or front and rear mowers. But the engineers at Winnipeg couldn't understand why I was not more enthusiastic about their baby, and insisted on one being put on the Ford New Holland stand at the 1987 Royal Smithfield Show, accompanied by an engineer from Winnipeg.

He was full of enthusiasm by the end of the event, and told me he had four firm orders, although we had not fixed a retail price. It's always amazing what people will say at a show if they have the attention of a representative. I found it hard to believe he could sell four and asked him for the details of the best potential customer, which turned out to be a big vegetable grower in Kent. I arranged for the tractor to be delivered to his farm for a two-week evaluation.

One of the key jobs he gave the Bi-Directional was loading potatoes into trucks, but unfortunately the loader's lift height and reach were insufficient, and later that week I had a phone call asking me to come and collect the tractor. The Winnipeg engineer found this hard to believe, but the farmer had made his decision without even knowing the price. The tractor was expensive in Canada, and we calculated that once rail transport to the east coast, shipping and import duty had been added, it would be possible to buy a Ford 7610 and a telehandler for less than a Bi-Di. I think that finally knocked the last nail in the coffin, and I was able to stop writing endless reports to HQ in America saying there was no market for these tractors in Europe.

It's not quite true to say we didn't sell any on this side of the Atlantic. A Swedish buyer purchased 25 units for a special application working in pine forests. The tractors were to be used in plantations to thin out every other tree and cut each into 2m long logs, which would be processed for pulp. Articulated steering made it possible to easily move between the rows of trees in the forest, and there was a lot of talk about another order for a further 25 units, but unfortunately it never materialised.

I must admit to winding up the North American product manager for the Bi-Di, who thought the tractor was the best thing since sliced bread, while I found it difficult to see a market for it in Europe. I even found an artist to draw a picture of a wheelbarrow with a wheel and handles at both ends. I had it framed and gave it to him, but he was not amused! Starting in 1989 we did, though, import the 325hp Ford Versatile 946, for customers who were keen to replace their FW tractors with another high-hp Ford articulated tractor.

***

It was at the same Smithfield Show in 1987, though, that we launched on to the market another tractor that would become a major success – although in this case it was designed and built in Britain.

Sometime in the mid-1980s I was in discussion with Eric May at South Essex Motors (SEM), to whom we contracted out special build projects. Our topic of conversation was whether we could create a compact 100hp six-cylinder tractor based around the 7610, to compete with others in this sector. After a lot of thought and planning, the SEM team got to work and built three prototypes. They were received with a lukewarm response by our engineers, who were unconvinced by the merits of a compact six-cylinder 100hp tractor, but SEM then suggested that it did the conversion and it would sell the tractors to our dealers.

At this point the purchase department got involved, and Geoff Tiplady was forbidden to talk to SEM – but no one told me not to. I had meetings with the SEM directors, who were worried they had already spent a lot of money on the project and did not know whether to continue. My proposal was that we could buy the drawings from them and build what would become the 7810 on the Basildon line.

That week, we had a meeting with our American vice president, at which I asked him if I could have a private talk on the subject. I showed him the pictures of the tractor and the parts drawings, and he agreed that Ford would buy the design from SEM. With help from Bob Smale in engineering, we were in production six months later. In its first year the 7810 made a profit of $6 million. The tractor went on to become one of Ford's most popular sellers of all time.

In 1989, the Basildon plant celebrated 25 years of tractor production, and to mark the occasion Ford released a limited number of 7810 'Silver Jubilee' tractors in a silver livery and with a high specification. We built 150 tractors, with the intention that each UK dealer would have one. Each was equipped with air conditioning, an automatic pick-up hitch, four spool valves, a performance monitor and many other goodies available at the time.

The problem was that this made them quite expensive, and combined with some scepticism about the silver paint, this led to a lack of enthusiasm among both dealers and farmers. As a result, some dealers in the arable areas, where high-spec tractors were in greater demand, took four or five, and some dealers in the west none at all.

Meanwhile, as a further measure to mark the factory's quarter-century, each farmer who purchased any Basildon-built tractor that year was given a pewter replica of the real machine he had bought. Both these models and the real 7810 Silver Jubilee tractors ultimately became very collectable.

The latter weren't the first of their type, though – some years earlier our dealer Hendry of Southampton celebrated its golden anniversary as a Ford dealer by painting a Fordson Major gold.

If someone still has it, then it will probably be worth at least as much as the prices now being paid for 7810 Silver Jubilee tractors.

12

Farmer Feedback

There were a lot of people who played a big part in my life with Ford, and a large number of them were true characters among our key farmer customers. Peter Philpot farmed close to Basildon – indeed he still does – and most people in the area know him either personally or by reputation. His farming venture extends to thousands of acres throughout Essex and Suffolk, and he was a big Ford tractor customer. Peter and I had one thing in common – we both had one O-level each! Today, I wouldn't even have got an interview at Basildon but in those days they took you at face value, and not on your paper qualifications.

I already knew him quite well when I bumped into him one evening during the 1980s at the local NFU dinner and dance at Ingatestone. He had acquired some big 2wd John Deeres to keep us on our toes, and couldn't resist the opportunity to tell me how good they were. At that time, conventional 4wd tractors were still relatively rare, although John Deere was one of the mainstream makers to first enter the market, albeit with a hydraulically-driven front axle, which tended to stall in heavy going, rather than the later, much more successful mechanical drive.

After coming to the UK market in the mid-1960s, Deere was relatively slow to establish itself here, but its tractors, coming from factories in America – where it competed strongly with International Harvester – and Germany, were proving reliable and well built.

Just after midnight, Peter suggested I meet him at his farm, Barleylands, at 6.30am the next day so he could show me why they were so good. No one refused an invitation from Peter!

I hardly had time to sleep and was outside the house as requested at 6.15am. Everything, however, was in darkness, but within minutes all was busy. Peter couldn't believe I had come. Jo, his wife, produced breakfast and we were soon roaring down to Potton Island in Peter's Range Rover. Once on the island I asked Peter what the red flag adjacent to the field meant. He suggested I take no notice, shortly before a huge explosion made me almost jump out of my seat. The army were apparently testing a new explosive! Shortly afterwards we arrived in the farmyard there to find both John Deeres out of action, one with a fault and the other with a rear puncture, so my chance to drive one didn't come on this occasion.

Whenever Peter had a problem he would ring me in the office at Basildon only a couple of miles away across the fields from Barleylands. I would often tell him to put the phone down because I could hear him shouting from where I was sitting, which always made him laugh. He was a bundle of energy as he strode out across the fields, and many people had to run to keep up with him. At harvest time he would aim to see nearly all his tractor and combine drivers every day, covering many miles between his farms.

Peter is an amazing man to deal with and incredibly generous. Between 1984 and 2004, I was invited to shoot with him at his Boyton Hall farm on 21 occasions. He never shot on his own shoot, as he had far more fun shouting abuse at any gun that missed a high pheasant – but he never seemed to notice a good shot!

One day I was talking to one of his tractor drivers, who had broken his leg and had to go to hospital. Peter visited him that night and asked why he didn't have a television. When he was told there wasn't one, Peter walked out, purchased a TV and installed it for the driver before he left. That sort of thing doesn't often happen these days.

When he said something I seldom disagreed, because I had to admit he was right 98 times out of 100. With this in mind, he was the ideal candidate to talk to our engineers. Back in the late 1980s a group of engine designers from the US were coming to Basildon to investigate the possibility of developing a new engine range, and wanted to talk to some farmer customers to find out what they wanted. Obviously, my first choice was to take them to meet Peter.

Few people with knowledge of tractors need reminding of the problems we had at times with our Basildon-built tractor engines. The biggest ongoing issues we had were with porous blocks, oil leaks and balancer gears. Reliability was a huge problem for the service department, but an even bigger one for the salesmen, who had to persuade customers that if they bought a new tractor they wouldn't have a repeat problem. We were eventually able to give an extended warranty on engines with porous blocks, and ultimately we decided to copy the engine water filtration that was fitted to engines such as the Cummins units used in the FW-30 and FW-60, which cured the fault.

We had a problem communicating issues to our engine engineers based in the US, particularly as the problems we had in the UK and Europe were far bigger than those in the States. I think some of this related to the far denser population of Ford tractors in the UK, while, in the US, both farms and Ford tractors were further apart. Tractors of under 100hp worked far harder in the UK and Europe than they did in North America, where they were used more as runabouts, leaving the big 100hp-plus machines that were still relatively rare over here to do the heavy cultivating.

In the mid-1980s, alongside planning for what would become the Series 40 replacements for the Series 10 tractors, development was beginning on an all-new range of four- and six-cylinder engines. It was code-named Genesis (not to be confused with the same name used in North America for the later 70 series tractors) after the first book in the New Testament, and would be the first completely new Ford tractor diesel engine since 1964, when the 6X range was introduced. It was a new beginning, and engineers from both sides of the Atlantic were to join forces and make it all happen.

About 20 engine engineers were coming to Basildon to identify issues and solve problems, and it was decided we would take them by coach to Barleylands to meet Peter and listen to what he had to say as he was such a hands-on farmer. I arrived at the farm ahead of the American entourage to find that Peter had brought tractors in from all over the county. There were nearly 30 Fords in a line, interspersed by half a dozen John Deeres, which were becoming strong competition.

Peter climbed into the coach and was introduced. Still in their seats, the engineers had no choice but to listen. He then waded in with both feet, starting by pointing out our engine oil leak issues and ending an hour later with the PTO oil seal leaks, missing nothing in between. Every other sentence was: 'Are you proud of this?' Not surprisingly, the engineers said nothing but squirmed in their seats. They had no idea we had such problems.

Eventually, they were allowed out of the coach and told to inspect each tractor. If any of them started to chat to each other they were told to pay attention. They were now all scared to death of Peter. We went to examine each machine, with Peter pointing to blackened cylinder blocks where oil leaks from cylinder head gaskets and rocker cover gaskets had burned the blue paint. We moved to crankshaft oil seal leaks, a problem that we had had for years. Each new seal was going to solve the problem, but it never did.

Then it was on to the hydraulics, and problems with spool valves dripping on to the yard. I had to admit to secretly being pleased we didn't make the spool valves ourselves. We moved to the back to inspect PTO oil seal leaks. Once again the engineers were asked loudly and forcefully: 'Are you proud of your tractors?'

We were then asked to look at the John Deeres, some with 7,000 or 8,000 hours on the clock. Upon examination, it was clear there were very few oil leaks at all. Even the head engineer had to admit there was a massive difference between the blue and the green.

Peter still had one ace up his sleeve and took a handful of spanners. He asked one of the engineers to drain the engine oil on one of the Fords and fit a new filter, knowing that when they removed the filter the old engine oil would pour down the side of the block. The engineer had to admit he had never changed a tractor oil filter in his life. Again the familiar phrase rang out. Our man wasn't proud of this, but he wasn't going to admit it now. Peter exploded, calling the engineers a bunch of ******* as he walked back to his office. A job well done!

It was an amazing meeting. The engineers had been crucified by a big customer of their products, with clear evidence of the issues involved. One thing was for sure though – they took notice of everything he said. The engineers returned to Basildon and remained closeted in the conference room for days on end, dissecting the feedback they had received from this and other visits, before packing their briefcases and returning home to the US. It really seemed as if, spending their lives in palatial offices, they were totally protected from the farmers and farming that ultimately paid their wages, and they really had no idea that we had so many problems.

***

The new Genesis range of four- and six-cylinder engines was to be built at Basildon, and would ultimately be fitted to the 75–120hp Series 40 tractors under development, codenamed P358, which would also be built there from 1991 onwards. In six-cylinder format they would also be used in the P396 tractors planned for launch ultimately as the 170–240hp Series 70, to be built at the Winnipeg plant in Canada from 1994. Somewhat confusingly, these latter tractors were also to be known as the Genesis range in North America, but over here were simply 'Series 70'.

The whole development experience was a turning point in the company's history as far as engine production went, to which we owed a lot to the response to feedback from customers such as Peter. For years it had been Ford policy for the engineers to build what they thought the farmers wanted, and on the sales side we got very little say in the matter until it was too late to suggest sensible changes. However, after the purchase of New Holland we began to take on some management ideas from the firm, which had a totally different policy, with dedicated product managers. We adopted this structure, and appointed separate tractor product managers for three-cylinder tractors up to 60hp, four/six-cylinder machines up to 120hp, and six-cylinder models of over 120hp. The idea was that the product manager told the engineers what the customers really wanted, and the engineers built it.

I had spent a large part of my career to date involved with big tractors, spending many hours working with the TW and FW ranges, and was delighted when my next appointment was that of product manager for our large tractor ranges, covering the whole world excluding the USA, which had its own manager. At the time, the P396 project to succeed the TW/Series 30 design was just getting under way.

We had seen American tractors come to the UK before, and I had spent some time testing the Ford FW machines in the US, but involvement in the P396 project gave me much greater experience of how different North American farming was to European. The object of the project was to create a much more 'worldwide' tractor at this power level, and I was spending a week a month in the Pennsylvanian town of New Holland, home to our new world headquarters since the Ford New Holland merger. While this was traditionally a baler and forage equipment factory, our global product engineering department was now also located there, so it was home to the initial stages of the new big tractor programme.

I loved the town of New Holland. It's the home of the Amish, a religious group originally persecuted in Germany who escaped to America and settled in Pennsylvania, where they made their home among some of the world's best farming land. Their farms consisted mostly of 300-acre blocks with 150 Friesian cows, but their religious beliefs meant they did not believe in using mechanical motive power. While a stationary tractor blowing maize into a silo might be an occasional sight, all field work was done with horses, and it was quite common to see an engine-driven New Holland baler being towed by a team of them. Road transport was by horse and buggy.

Once a week there would be a farm equipment sale in town, often an unbelievable sight. Binders, horse ploughs, 8ft rolls, butter churners, horse harnesses – it was like a farm sale in 1930s England. The people were always very friendly and wanted to know how we farmed in the UK.

They did not have phones in their houses but there tended to be a phone box in the middle of a field for emergencies only. If you were seen to spend any time on the phone, everyone knew about it. The Amish also had a wonderful housing system. Mother and father built a new home joined to the house of their mother and father, and most houses were home to three generations. Life, however, was very strict, with no alcohol allowed, no cars and no girl or boyfriends for the youngsters. When the children were 18 or 19 they were allowed time out, and those local to us tended to visit a quarry where they kept motorbikes, drank beer and did all the things teenagers do! By the time they were 20 they were expected to buckle down and work having got all the fun and games out of their systems. Very few ever left the family farm, and most of the Amish earned good money from their wonderful red dirt soils.

Many of the bigger local farms farmed only 50 per cent of each field each year, fallowing the remainder, and the major work was done by big, four-wheel drive articulated or tracked tractors. Fields would be divided into strips about 60 yards wide, with the preceding year's stubble left untouched on the intermediate strips to stop wind erosion. Those in between would be drilled with spring cereals or maize. With this system and these machines, hundreds of acres of work would be done in a day, using systems based mainly on minimum tillage 3 or 4in deep, with very little ploughing.

The big dual-wheeled articulated tractors would be driven mainly by Mexicans, and would cover thousands of acres with seed and fertiliser in a very short time. But there were also plenty of TWs at work, although operated very differently to how we would expect in Europe's heavy clays. Where we tended to overload our tractors working in lower gears, the Americans preferred to work more shallowly at higher speeds, with the tractors less heavily laden. You would never see a TW struggling with a five- or six-furrow fully-mounted plough in heavy clay. Indeed, the plough was a fairly rare sight, with just a few semi-mounted units around, the main work being done by lots of tined cultivators and big sets of disc harrows.

One of the issues Ford aimed to address with the P396 development to replace the Series 30/TW was robustness. In England we were having endless troubles with crown wheels and pinions and rear transmissions in both the TW-25/8730 and TW-35/8830, as well as breakages of lower links, drop arms, top links and flex ends. I accompanied a group of engineers who came to England to study the rear transmission issues in particular, and we visited farmers across East Anglia, Lincolnshire and Yorkshire. Many were using heavy five- and six-furrow fully-mounted Dowdeswell reversibles. A solution was required to the stress issues these were creating, but the proposal from the engineers was to put a disclaimer in the front of the operator's manuals saying that ploughs over a certain weight would invalidate the tractor's warranty! I was not impressed, only imagining how Case IH and John Deere would delight in using this against us.

Eventually the engineers returned to the US, but not long after that I was over there myself, attending a meeting along with a number of others with Bob Moglia, the worldwide boss of Ford's tractor business. During the course of the meeting, the engineers gave their report on their TW studies in the UK and elsewhere, but no mention was made of any problem with TW hydraulics. I questioned this, but the section of the engineers' report pertaining to hydraulics was again read out, and it completely contradicted the facts relating to Europe. I realised that, believing top management did not like bad feedback, the engineers made sure they only got good news.

All that night I kept thinking of the fraud that was going on before my eyes, and made up my mind to rectify the situation the next day. I went to the top management office suite, spoke to Mr Moglia's secretary, and asked for a private 'off the record' meeting. She spoke to her boss and I was told to return at 11am for 15 minutes.

I arrived on time and told Mr Moglia I wanted to talk about the TW hydraulic and linkage issues that had been discussed at yesterday's meeting. I didn't mince my words, saying that what he had been told about TW performance and reliability in Europe bore little relation to the truth, and that we had major problems with lift capacity and linkage failures. Just over an hour later the meeting was over, Bob declaring his regret that he had not listened before to what I had been trying to say.

We walked together into a meeting room in which the top engineers were in conference. Bob was a little guy, but he got people's attention very quickly. It was perfectly obvious that I'd blown the whistle, and things went very quiet. Bob asked for a complete update on TW hydraulics and rear transmissions. I disappeared before I got lynched ...

13

New Tractors, New Owners

The original plan for the P396/Series 70 tractors to replace the Series 30 called for six tractors ranging from 135–285hp. However, amid all this came the news in 1990 that Ford had succeeded in doing what it had been planning since acquiring New Holland and Versatile to form a full-line, worldwide farm equipment maker. It had found a buyer for its farm machinery interests, allowing it to focus on its core car business. Our new parent was Fiat, the Italian conglomerate that had its own interests in farm equipment, as well as cars, trucks, construction equipment and many other areas.

As it became clear how its management wanted to combine the two organisations, working on a common platform while retaining both colours/brands, we made many trips to Fiat's tractor HQ in Modena where we worked on projected volumes for both blue Fords and terracotta Fiats. One of the key reasons for this approach was that our markets were very different, and very seldom did we compete in the field, with Fiat being strong in the southern hemisphere and Ford in the north. Fiat sold lots of tractors in Latin America, southern Europe, Africa, the Middle East and, to an extent, in Australia and New Zealand.

Production-wise, Fiat's business also operated differently, with sub-assemblies, engines, gearboxes and rear transmissions being assembled and then coming together at the end of the line. At Basildon, production started with a gearbox and rear transmission, before the engine and then other key components were added as the tractors moved down the line.

A new parent meant new hurdles to overcome in terms of convincing a different level of top management to fund the project. We needed $6 million to finance the new range, and spent a number of weeks during 1991 preparing for a presentation in front of our new Turin-based owners.

We knew the TW-based Series 30 range just could not compete with the more powerful ranges offered at the time by John Deere, with its 55 series and imminent 8000 range, and Case IH, with its Magnum tractors. I produced figures to show the European market growth of the 180–240hp sector in which, unlike John Deere and Case IH, we weren't yet competing. Outside of North America, the big markets for tractors in this power bracket were in France, Germany, Holland, Australia, New Zealand – and the UK.

This was all on the screen, and it was clear that Ford's tractor business, for which Fiat had just paid an awful lot of money, was missing out on a significant chunk of a growing market. The top Fiat man present suddenly stood up and without further hesitation declared that there wasn't a decision to make, and we were to be given the money we needed.

The new combined Fiat/Ford New Holland business, which was to use the name New Holland as a neutral and widely recognised brand, continued to have its North American engineering HQ in the picture postcard town of New Holland, Pennsylvania, and it was here that engineers from around the world met to plan the next stage in the development of our new big tractor range. The chief engineer was a very young chap, Dave Templeton, who was a brilliant engineer and very soon wanted to know everything different markets around the world required.

In North America and Australia, it was dual rear wheel mounting via extended axles, while European farmers wanted their tractors to run efficiently on big singles. In Europe we also needed sufficient lift capacity to hoist heavy 6m power harrows and big six- or seven-furrow reversible ploughs, so required hydraulic lift rams twice the size of those on US-spec tractors, a large proportion of which would actually be specified in drawbar-only format anyway. We even identified a need for a totally different sequence of gears in the powershift transmission, the US 'box having different speeds and ratios. One ingenious item of specification that was primarily targeted at the US rowcrop market but which also gained some uptake here, though, was the SuperSteer pivoting front axle.

The tractors were mainly tested internally on test rigs, but we had half a dozen in the field during the early 1990s, mainly in Texas and California, and I spent a lot of time testing them in the latter, ploughing in lettuces after the pickers had taken all the best ones. I wasn't entirely happy about the lift capacity on the 70 series and was determined to make sure the engineers had given us what they had promised.

We managed to find a big Kverneland plough through a local dealer, who agreed to lend it to us. It was only a four-furrow reversible, but was a heavy plough with extra-long legs providing a huge clearance, and we hooked it up to a Series 70 prototype. I ploughed acres of land with it, and while the land we were on was beautiful soil that went down metres deep with not a stone in sight, it was obvious very quickly that engineering had been true to their word. As a further test, I drove it flat out down the farm roads, but the linkage survived. I do recall one day, though, where the tractor was given an unintended highway test. I was on my way home from the field when I found myself taking a wrong turn and driving along an interstate highway, the equivalent of our motorway. It was illegal to take a tractor on these roads, and it was with some nervousness that I found myself driving nearly 20 miles before I could get off. That was probably the first and last time a Ford Series 70 tractor took a trip down the interstate.

I was really pleased with how the engineers who were developing the Series 70 tractors had listened to our needs. On another day I was working with a five-leg subsoiler, an excellent way of testing the draft control. However, at this early prototype stage it wasn't working correctly, with the subsoiler rising and falling in the ground and unable to keep an accurate depth. I telephoned the chief engineer at HQ in New Holland and he arrived by plane early the next day. Although he was very concerned and thought I was imagining things, he soon realised there was a problem and by lunchtime we were back in the field and the hydraulics were working perfectly.

While doing this field work with the Series 70 in California, I met a lot of Mexicans working on the thousands of acres of lettuces there. They would be manning 15 row harvesters, taking each head of lettuce from a conveyor belt before placing it in a plastic bag, with boxes of heads then placed on a trailer. One of the Mexican tractor drivers told me he was unable to get a work permit to stay overnight in the US, and as a result drove 100 miles to and from work each day. Providing he got just one hour's work in America it was equal to ten hours' work in Mexico – and, of course, the fuel over there was very cheap.

Often we were working in fields surrounded by apricot trees, with the fruits almost – but not quite – ripe. One of the Mexicans told me to cut some in half and leave them in the hot sun, and then have them for lunch the next day. They were delicious, and I lived off apricots all that week. At lunchtime a small white van would arrive selling Mexican wraps, and it was suggested I try one. It was so full of chillies it nearly blew my head off, and after that I went back to lunching on delicious sun-dried apricots.

We needed a range of land and job types over which to test the prototype tractors, and next did a deal with a sod farmer to carry out some operations on his land. While at first it seemed an impolite term, all quickly became clear when the nature of his crop was revealed. Sod is the American word for turf, and he produced this for the gardens and sports surfaces of California.

There were 200 acres of perfect soil that, it was believed, went down to 18ft. Removal of each sod, or turf, took about half an inch of topsoil with it, so there was plenty available for many years to come. As soon as the turf was cut and carted away on pallets on massive lorries, the land was then reworked for the next crop of turf.

Our first task was to subsoil the land to a depth of 22in to get rid of the surface material, before it was disc harrowed using two sets in tandem with a rear levelling bar. This was then followed by a rotary cultivator driven slowly through the ground, which could dry out quickly in the blazing sun. The land was then ready for re-seeding with top quality grass seed, but before this began, very fine netting was put down to help prevent the sods from breaking up when lifted. The netting was held down by what looked like hairpins to ensure it formed the base of the sod.

They then introduced a quad bike to sow the grass seed into the prepared land, after which the next operation was to lay out the irrigation equipment, with pipes set up every 25 yards and irrigators every 25ft, throwing out a very fine mist of water to get the grass seed to chit. The irrigators would be moved every few hours by a man on a quad bike with a very long stick. He would drive down the rows hooking his stick around the pipe and moving it over a few feet.

In no time at all the new grass would be up. As soon as the sods were ready to cut, the grass was mowed to a height of 1in by a rotary mower that was attached to the front of the sod cutter. This cut individual sods before rolling them up ready for transport to their new destination. The sods were stored on pallets on the cutting tractor, and then removed and transferred to a truck.

As big tractor product manager for the world excluding North America, there were the needs of several other markets beyond Europe to take into consideration, among them Australia, New Zealand, France, Germany, Holland and Belgium. I knew nothing about Australia, so in the next part of my work on Series 70 development I packed my bags and flew to Sydney, where I was met by one of the Australian zone managers, who looked after me very well.

We assessed the size of the market for 300hp-plus tractors, and I quickly realised that everything was big in Australia, particularly the fields. Not only was there a good market for the size of tractor on which we were working, but there was also strong demand for high-hp articulated tractors.

The first thing on our agenda was to visit Australia's equivalent of the Royal Show, which was quite an eye-opener. The extent of the equipment I had never seen before again highlighted how different was the scale of the country's agriculture. Then there was the specialist machinery, such as land-levellers to grade fields and help prevent water loss. It was a fabulous trip filled with lovely people, and going to Australia gave me a picture of how different the country was to what I was used to in Europe, helping me factor the needs of Australian farmers into the plans for the Series 70 tractors.

While our Antwerp factory in Belgium had been responsible for the production of TW/Series 30 tractors, Series 70 manufacturing was planned for the Versatile plant in Winnipeg, Canada, where there was spare capacity alongside the production of articulated tractors. The Antwerp facility would revert to the production of components. Originally, the plan was for a six-model range, but it was soon obvious that budgets would restrict that to four. Helping on the engineering side was a lovely man called Glen Karley, who had spent most of his life with John Deere, and more recently had been with Caterpillar. He was a collector of International Harvester pick-up trucks, and whenever he would see one on the road would follow it home and offer to buy it on the spot.

I spent a week a month in HQ at New Holland over the course of the development programme, and by my calculations flew the

Atlantic at least 110 times, getting to know the airport ticketing staff very well. By now all air travel was economy unless you were flying for longer than ten hours, but I used air mile bonuses and regularly got upgraded to business first. I tended to use Continental Airlines, flying out of Gatwick, which was much easier than Heathrow. Coming back from Newark, New Jersey, one of the ticket staff I got to know had a young son who loved model tractors and diggers, and we had a trade whereby I would find him toys to take home and he would upgrade me from economy to business!

With development well under way, we spent long periods of time working on the volumes of parts needed to build every tractor, starting with an estimate of what would be needed for each model – although they were essentially very similar – and then working our way through the options likely to be required in key markets, including lift cylinders, SuperSteer axles, front weights by volume, front weights by size, front and rear wheel sizes and 2wd and 4wd front axles – there was still a strong North American market at the time for 2wd machines of this size. Then there were the different market demands for wide axles, drawbars, rear linkage quick couplers or fixed ball ends and numbers of spool valves.

It was vital that the purchase department knew what to order, so that parts were available on the line when the tractors were built. Eventually, though, all planning was complete and production was ready to commence in 1994.

We held a worldwide launch for the Series 70 tractors in London, hosting the largest number of people we had ever seated for lunch at one time. For the UK market, this was followed by a series of events up and down this country, and we were very pleased with the response from large tractor operators, who were delighted with the clean-sheet design and the fact we now had a top model with almost 55hp more than we had previously had in a tractor of this format. Indeed, the new range began with a tractor that produced only 16hp less than the 8830.

I helped launch the Series 70 tractors in a number of markets, and particularly recall events in New Zealand, where I met up with my very good friend John Parfitt, who managed the country for the company. We spent the first two days at the country's national agricultural show, where we had been given a fairly large plot enabling me to show off the benefits of the powershift transmission and SuperSteer front axle. We then moved on to demonstrate a Series 70 on both the South and North Islands, and I particularly recall visiting a lot of vegetable growers with 100-acre fields in the north of North Island. At the evening gatherings after each day's demonstrations, I have hazy memories of being invited to join farmers who had a beer mug in the right hand and a jug of beer in the left, topping up the mug at regular intervals. I soon found myself holding the same – when in Rome, do as the Romans do! Of all my overseas trips, that two weeks in New Zealand was one of the most memorable in my 40 years with Ford.

I was very proud to be so closely involved with the Series 70 development, a range of tractors built to my own shopping list to take the place of the outdated TWs. When Fiat decided to purchase Case Corporation and merge it with New Holland, it was unfortunate that the Versatile factory in Winnipeg and the product lines that went with it had to be sold to meet monopoly rules. The view of the competition authorities was that two articulated tractor lines – Case IH Steiger and New Holland Versatile – and two large conventional tractor ranges – Case IH Magnum and the by-then-updated New Holland 70A – would give the merged business too dominant a position in these power brackets. Production of the Series 70A tractors and later derivatives continued well into the 2000s, though, under the colours and name of new owners Bühler Versatile.

14

Car Connections

Whether as Ford or New Holland, ultimately Basildon has always been owned by a car manufacturer. Working for Ford Motor Company had big benefits when it came to cars. Sometimes we were given a choice of what we could have, but when a car wasn't selling well, we had to take what we were given.

Upon my return in 1959 from my first overseas trip to Central and South America, to demonstrate the Dexta, I was able to order a new car, a Ford Anglia 100E. It cost £440, and it was about four months before I could take delivery, but when I finally collected it I was a very proud man. I had to keep it for a year, and at the end of that period I called in at the Ford car dealer Gilbert Rice on my way home to Sussex and they offered me £500 against a new one. That was the only time I have ever made a profit on a new car.

When, in the 1960s, I was promoted to area manager, I had a succession of Mark 1 Consuls. My next car was the Ford Classic, the one with four headlights. We were often given cars on a test basis about a month ahead of their launch. The Classic had a 1500cc engine but was a bit lacking in power.

Later, there came a time when Ford Zephyrs were not selling well and we were continuously getting new ones, some with engine numbers older than the one we were driving. I must have had five or six of these, which were sluggish to say the least. The company had just announced a new Anglia, which despite being very small was very quick off the mark. Deciding I needed something a bit quicker, I asked if I could have one, but soon found that if you were at a T-junction wanting to turn right, the front wheels would spin on take-off and you would stand still. After about four months I decided it was too good for me and asked for another car instead.

At this point, I was looking after two areas between Kent and Cornwall, covering around 55,000 miles annually. One day when I was just five minutes from home there was a huge bang from the engine but I was able to nurse the car home before investigating. Upon taking it to the local Ford car dealer I discovered that one of the pistons had broken in half!

By the time I was southern area sales manager, Ford had recently introduced the Lotus Cortina, and after a lot of discussion, my boss Laurie Chivers agreed that northern area manager Arthur Hebblethwaite and I could both order one. They were fantastic cars, being small but very powerful, and could reach a steady cruising speed of 100mph without a problem. Because of this, I had a few exciting moments in two of the ones I had.

On one occasion, I was on the Witham bypass near Basildon, having a race with a sports car. This, though, was the point at which the engine suddenly decided it had had enough of being revved to reach 100mph, and it promptly exploded. Luckily I knew that Doe Motors, the car arm of the tractor dealership, was located not far from the next turn-off, and I was able to coast the last half mile to their workshop door, where I left the poor Lotus Cortina to be looked at.

If I was going to visit a dealer in the West Country, I used to leave home at 4.30am, aiming to be at the dealership by around 9am. I was often asked where I had stayed the night by dealership principals, who couldn't believe I had driven straight from home. I did have five Lotus Cortinas in succession that were very quick, and no one had yet invented speed cameras. One day I was running late on my way to see Mr Slade, the MD of the Rank dealer organisation in Devon, and driving very fast. There was a humpback bridge about 50 yards ahead of me, and 50 yards later I had all four wheels off the ground. To my horror I then saw not far in front of me a cattle truck that I had assumed was doing between 40 and 50mph. However, I quickly discovered it was not a relatively rapid cattle truck, but a cattle trailer being pulled by a Massey Ferguson 35 at about 15mph. Somehow, and I will never know how, I avoided hitting it and ending up in the back with the cows.

One night I was attending a farmers evening at Watson and Haig, our dealer in Andover, Hampshire, which ended at about 9.30pm. Subsequently, I had to drive back to Essex and then get up early for a dealer meeting in Diss, Norfolk, the next day, so got into my Lotus Cortina and pushed on home.

Not far out of Andover, there was nothing on the dual carriage-way and I was doing about 100mph for a good ten miles, but had to slow down when faced with a 40mph limit. It was then that I noticed a car coming after me very fast, with flashing blue lights. I pulled over and awaited the tap on the window from the police-man. I explained my situation, but he informed me he had been trying to catch me for ten miles in his Austin Westminster. How-ever, he must have been having a good evening because, although I then made my second mistake by telling him he would never have caught me if I had known who it was, he then admitted he had enjoyed the chase, and I was fined just £10 – cheap at the time!

In the 1970s, Sir Leonard Crosland, chairman of Ford Motor Company, bought a 600-acre farm in Great Wigborough, Essex, with which I became very involved. I was meeting Geoff Tiplady at Basildon one day when I was informed of the purchase and given the task of equipping the farm with a suitable fleet of machinery.

I met Sir Leonard at his house in Kelvedon and we drove in his Lotus Cortina the ten miles or so to the farm, which we toured while he explained his improvement plans, such as installation of concrete roads. Once we had decided on the equipment he needed we drove home via the old Boreham Airfield. When we got to its main gate we met the security guard on duty who, despite Sir Leonard's suspicion that he would not be recognised, immediately waved us on through the airfield test track. After driving around the track he headed towards a man-made hill used for truck testing. We hit the top at nearly 90mph and all I saw was sky. All four wheels were off the ground and I was utterly terrified, but we managed to land safely on the downslope.

Despite all that fun in Ford cars of the 1960s, though, I have to admit that my favourite car was the Capri, of which I had at least five. Initially we were given standard 1600cc models, but later during the 1970s came ones with three-litre engines, fuel injection and four-wheel drive.

In 1991, when Ford agreed to sell its farm equipment operations to Fiat, it resulted in a change of the cars available to us. As Alfa Romeo was part of Fiat, their cars were one option, and I loved them. I was lucky enough to be able to run six before I retired in 1994, and still have one today.

15

The Future – Then and Now

Back in the mid-1970s I took a phone call from a man who said he would like to bring two colleagues and visit me at Basildon. I asked what he wanted to discuss, but he said he would prefer to wait until we met. Intrigued, I invited them over to lunch the following week, after which I suggested I could give them some time. They arrived at Basildon at 1 o'clock, and it was immediately obvious they were not farmers. I had no idea what was coming next.

Back in my office, we sat around my desk and they produced a mass of drawings. They said they had developed a new tractor with a high road speed, which they were calling the Trantor. They said they believed it could take over the role of smaller tractors on many farms, and replace the role of traditional tractors on hill farms, where draft work was not a requirement. Then they asked if Ford Motor Company would like to opportunity to build and market it.

We looked at the drawings and it soon became clear to me that it was not unlike a Land Rover, but fitted with big rear wheels and a pto and hydraulic linkage. They suggested it would be easily capable of cultivation work, as well as doing all of the large element of road work that makes up much of a tractor's time, but in this case at speeds of up to 50mph. I was imagining how the Trantor could plough with its small wheels, and trying to work out whether it could have a loader fitted, a prerequisite for smaller livestock/mixed farm tractors in those days. I could see it doing everything they said it would, but all of these jobs could be done by a second-hand Dexta for far less money.

They, however, were convinced that if the Trantor went into production with someone else it would cost both Ford and Massey Ferguson a lot of their small tractor business. They continued to sing its praises, suggesting Ford would be making a big mistake if it turned this opportunity down.

I found it very hard to see a big market, and dearly wanted to tell them that it really did not have a future. I didn't want to upset them too much, though, and explained that we didn't have the space at Basildon to take it on. I also warned them that it might be risky to invest their own hard-earned money in such a project. My knowledge of farming and farmers told me that multi-purpose tools were never a great success, and often proved to be or were seen as jacks of all trades and masters of none. They were upset about my lack of interest, but at least I had been honest with them.

They did, in fact, manage to put the Trantor into production for a period during the 1970s and early 1980s, and I often saw one pulling a trailer-load of straw bales when travelling to work at Basildon, although it did not reach great heights of commercial success. Ultimately, after some UK units were built and sold, a new production home and market was found in India, where it struck me that there could well be a good market and where manufacturing costs would be lower.

It was not long after this that I received a phone call from a farmer in Oxfordshire, who suggested he had something I would like to look at. I met up with my good friend Malcolm Denton, who lived in the area, and very soon we were looking at a prototype JCB Fastrac that had been loaned to farmers for evaluation. The tractor had an excellent mid-mounted cab with good fore and aft vision. Very soon we had it out on the dual carriageway with a wheat-laden trailer. I had never driven a tractor at more than 20mph, and to soon be doing 50mph was a great thrill. I had to admit it was very stable even flat out. The machine was obviously the result of a great deal of research and investment.

I still believe the Fastrac to be a specialist machine with its own market, and I don't remember ever being on a deal where a farmer was considering one of our more conventional tractors against a Fastrac. Soon after this trip to Oxford I was visiting Eric May at his workshop at SEM when I realised I was leaning on the front of a Fastrac. Eric was helping with the design of the gearbox, but he never told me about it until it was commercially available.

There was, of course, another high-speed tractor on the market, the Mercedes-Benz MB-trac, capable of 25mph at a time during the 1980s when most other tractors on the market could manage only 20mph at most. We had a customer in Norfolk who had a fleet of Fords and a number of MB-tracs, and one day he offered to show me what one of the German tractors could do.

It was actually his daughter who was going to be my chauffeur, and she jumped up into the cab, backed up to a big trailer, and we were off up the road on to the dual carriageway. Very soon we had hit top speed, which seemed to be a fair bit more than 25mph. I was especially impressed when, coming up behind a truck fully loaded with ballast, she pulled out and overtook. The driver did not look impressed! While it was well-conceived, the MB-trac was another low-volume seller, though, and even though they had a very dedicated group of customers, Mercedes-Benz eventually stopped MB-trac production. Perhaps these stories vindicate our decision not to get involved in the high-speed tractor business, particularly as modern developments have brought us cab and axle suspension and top speeds of over 30mph on conventional tractors.

By 1994 I had spent 40 years with Ford – working hard, living through lots of problems, but enjoying lots of great times as well. Along with many friends, I had a lot of fun, and I wouldn't have wanted it any other way. There are almost too many names to mention, particularly among our dealers – Roger Barclay, Eric

Bell, Francis Disney, Alan and Colin Doe, Chris Green, Roddy Robertson, Chris Rook, Paddy Campbell, John Hart, Peter Scott, Bob Webster, Dick Wilder ... I also made friends with some wonderful farmers, with too many names to mention here.

I retired in the middle of that year, just at the point Fiat was starting to create the New Holland brand, so putting a neat end to my career as a Ford man. I didn't particularly want to, but Dick Kiefer, the man who had become my boss but who I had never met, rang me from Fiat's London offices and asked if I would take an attractive deal for early retirement. However, it wasn't long after that when he made contact again and asked if I would come back for two years under contract. I willingly agreed, and shortly afterwards received a five per cent pay award. Then, to my total surprise they started paying my tax for me as well. All's well that ends well.

Upon my departure I was touched to receive a tribute from Bob Leary, general manager of sales and marketing in the UK for Ford Motor Company, even though the company was by then no longer officially involved at Basildon.

He said: 'You have made an enormous contribution to Ford in the United Kingdom. Probably no one else is better known and thought so highly of by our customers.'

I have only one regret from my time working in tractors for Ford, and that was that I was perhaps born 15 years too early. All I ever wanted was to be able to introduce more power to the tractors to boost the efficiency of farmers and farming. When I began working with Ford, we had just the 60hp Major, followed shortly afterwards by the 32hp Dexta. These machines were relatively cheap, and labour was plentiful, which meant many farms were able to run multiple tractors, with a UK market totalling around 35,000 units. In this period after the Second World War mechanised farming was rapidly replacing horses, while there was no shortage of tractor drivers, and so little need for bigger tractors.

By the early 1960s, though, labour was plentiful and drivers were poorly paid, exposed at work to cold and wet, dust and heat, and living in poorly maintained tied cottages. It wasn't long before many started to desert farming as a career for better prospects and conditions working in factories and offices. This labour shortage is why I wish I had been able to help to more quickly introduce more power.

When I look at what engineers have created today, with tractors twice as big as an FW-60 and with five times more power than a 7000, I can only look on in wonder. The demand for more power continues to grow, though, with tractors of 400–500hp now not uncommon not just in North America, but in the UK and parts of mainland Europe. I would love to have had the opportunity to sell the big New Holland T9 and Case IH Quadtrac tractors of today – with almost 700hp they are over ten times more powerful than the Fordson Major I was brought up with. I can recall talking at Smithfield one year with a farmer who was bemoaning the price of machinery, and him saying that if tractors ever cost more than £1,000 and combines £3,000 the industry would find orders drying up. With top-end tractors and combines today listed at over £100,000 and £300,000 respectively, his prediction doesn't seem to have come true.

Much of this is driven by the very high cost of labour. Good tractor drivers are increasingly few and far between, and long hours are still a prerequisite at busy times such as drilling and harvest. Many are often still working 16 hours a day during these periods to take home what those in other trades can earn in eight hours on a five-day week. I know of one tractor driver with a young family who pointed out to me that they were in bed when he left home in the morning and asleep long before he got home at night. He was already attending night school and learning to be an electrician, and he's likely to not be the only one with this idea.

***

Now, though, I had to find new challenges to occupy me in retirement. When I was 18, before I joined Ford Motor Co, I had been a fairly good golfer, playing off single figures, but I gave it up as even on my days off I just didn't have time to devote a whole morning to a game. So 40 years later I took it up again and was soon playing three times a week.

I had been so busy all my life that I soon started looking for something else to do. A friend who had a small shoot on his farm lost his gamekeeper and asked me if I would like the job. I had always loved the countryside and had kept ferrets, which I worked with nets. I taught many youngsters who had the patience and enjoyed digging for a ferret that killed underground. Being a gamekeeper involved the purchase of 1,500 poults at six weeks old, releasing them in a pen and then feeding them on a daily basis through to mid-January, while doing my best to shoot the foxes that enjoyed a pheasant when they got the chance.

During the shooting season I had to get together a team of beaters, who did the job because they enjoyed it. It certainly wasn't for the money! We had about 20 acres of maize outside our main wood, and drove the pheasants back into it, hoping the guns would shoot them on the way out. Shooting days were a great worry to me, as I could never be sure that the pheasants would be where I expected them to be. The guns never worried though! I enjoyed so much being with like-minded countrymen and I did the job for four years before handing it on to a friend.

I was also able to spend more time with my wife, Mollie, who I met when I was working as a demonstrator, and she was assistant housekeeper at Boreham House. To celebrate our golden wedding anniversary, I decided we should hold our party there. By that time it was no longer part of Ford Motor Co., and had become a wedding and conference venue. I rang the lady owner and she invited me to come and meet her, and she couldn't have been more helpful. The local butcher produced some very special beef and a friend provided the wine and champagne. We had an excellent party at the place where we first met more than 50 years ago, not long after my career with Ford had begun.

My advice to anyone reading this would be to make sure you plan your retirement long before it happens and if possible work a three-day week with a four-day weekend before retiring. After a long career working in farming, it never leaves you.

***